Publications Handbook
and Style Manual

1984

AMERICAN SOCIETY OF AGRONOMY
CROP SCIENCE SOCIETY OF AMERICA
SOIL SCIENCE SOCIETY OF AMERICA
677 South Segoe Road • Madison WI • 53711

ACKNOWLEDGMENT

Many persons contributed to the preparation of this *Publications Handbook and Style Manual*. Headquarters office prepared the final version after several drafts by the committee: D.R. Buxton, chair; R.C. Dinauer, D.A. Fuccillo, J.J. Mortvedt, and C.O. Qualset. Members of several editorial boards, officers, and members of the societies, as well as societies headquarters office staff, also contributed suggestions. Ann M. Diliberti assisted by copyediting the manual. Julia L. McDermott designed the cover.

Copyright © 1984 by the American Society of Agronomy, Inc.
Crop Science Society of America, Inc.
Soil Science Society of America, Inc.

ALL RIGHTS RESERVED UNDER THE U.S. COPYRIGHT LAW OF 1978 (P.L. 94-553)

Any and all uses beyond the "fair use" provision of the law require written permission from the publishers and/or author(s); not applicable to contributions prepared by offices or employees of the U.S. Government as part of their official duties.

American Society of Agronomy, Inc.
Crop Science Society of America, Inc.
Soil Science Society of America, Inc.
677 South Segoe Road, Madison, Wisconsin 53711 USA

Library of Congress Cataloging in Publication Data

Publications handbook and style manual.

   Bibliography
   Includes index.
   1. Communication in agriculture—Handbooks, manuals, etc. 2. Agriculture—Authorship—Handbooks, manuals, etc. I. American Society of Agronomy.
S494.5.C6P83   1984   808'.06663021   84-20460
ISBN 0-89118-084-2

Printed in the United States of America

Contents

CHAPTER 1 INTRODUCTION 1

CHAPTER 2 JOURNAL MANAGEMENT AND PROCEDURES 3

    Eligibility of Authors, 3
    Manuscript Handling, 3
    Publication Charges, 4
    Prior Publication, 4
    Notes, 4
    Letters to the Editor, 4
    Editor-in-Chief, 5
    Other Editors, 5
    Agronomy Journal, 5
    Journal of Agronomic Education, 8
    Journal of Environmental Quality, 8
    Crop Science, 9
    Soil Science Society of America Journal, 10

CHAPTER 3 PROCEDURES FOR *CROPS AND SOILS MAGAZINE* 11

    The Editorial Board, 11
    Types of Articles, 11
    Article Format, 12
    How to Submit Material, 12
    Length of Articles, 12
    Tables, Illustrations, and Photographs, 13
    Other Submission Pointers, 13
    Review, Editing, and Rewriting, 13

CHAPTER 4 PROCEDURES FOR MONOGRAPHS, BOOKS, AND OTHER PUBLICATIONS 15

    Monographs, 15
    Books, 18
    Special Publications, 18
    Agronomy News, 20

CHAPTER 5   PREPARING THE MANUSCRIPT............................21

   Manuscript Format for Journal Articles, 21
   Manuscripts Submitted for Other Society Publications, 26
   Details of Manuscript Preparation, 27

CHAPTER 6   CONVENTIONS AND STYLE..................................35

   Nomenclature and Terminology, 35
   Specialized Terminology, 37
   Statistical Analysis and Experimental Design, 39
   Measurements, SI System, 41
   Abbreviations and Symbols, 52
   Spelling and Capitalization, 54
   Numerals, 55
   Punctuation, 55
   Compound Words and Derivatives (Use of Hyphen), 56
   Time and Dates, 57
   Miscellaneous Style Points, 57

CHAPTER 7   TABLES, ILLUSTRATIONS, AND
MATHEMATICS 61

   Tables, 61
   Figures (Illustrations), 62
   Photographs (Shaded Material), 63
   Graphs and Charts (Line Drawings), 65
   Preparing the Drawing, 66
   Mathematical Equations, 71

CHAPTER 8   PROOFREADING..75

CHAPTER 9   COPYRIGHT AND PERMISSION TO PRINT..........79

CHAPTER 10   PUBLICATION TITLE ABBREVIATIONS.............81

REFERENCES..85

OTHER USEFUL LITERATURE 86

INDEX 87

CHAPTER 1

Introduction

Dissemination of information is one of the primary functions of the American Society of Agronomy, Crop Science Society of America, and Soil Science Society of America.

Publications from the three societies cover the following areas: (i) original research reports in agronomy, crop science, soil science, environmental quality, and agronomic education; (ii) reviews of research in these fields; (iii) symposia proceedings; (iv) monographs and textbooks; (v) abstracts of papers presented at the annual meetings; (vi) official business of the societies; (vii) news, events, and announcements; and (viii) popular articles.

The general editorial policies and practices of the societies are determined by the respective boards of directors, whose members are elected. Responsibility for maintaining editorial standards and advising on publication policy is delegated to editors-in-chief, editorial boards, and editors. A separate committee coordinates editorial policies and procedures among the societies. The societies' headquarters office staff manages the publications.

The *Publications Handbook and Style Manual* serves as a guide for authors in preparing manuscripts and other material submitted for publication by the societies. It replaces an earlier manual published in 1976, and it should be used as a primary source for writing, style, editing, and procedures for publications of the associated societies. Other books, e.g., the *CBE Style Manual* (Council of Biology Editors, 1983) and *The Chicago Manual of Style* (University of Chicago Press, 1982), may be used as supplements and expansions on these subjects.

CHAPTER 2

Journal Management and Procedures

Much of the administration and all of the printing, production, and distribution of the publications are coordinated by the headquarters office staff. This section gives an overview of the present responsibilities and practices in the review, handling, and production of publications of ASA, CSSA, and SSSA. Additional details and supplementary information are available from the headquarters office.

ELIGIBILITY OF AUTHORS

At least one author of each unsolicited manuscript published by the *Agronomy Journal, Crop Science,* or *SSSA Journal* must be an active, emeritus, or graduate student member of one or more of the societies. Exceptions to this rule may be granted by the governing body of the society. Authors of manuscripts published by the *Journal of Environmental Quality* or *Journal of Agronomic Education* do not need to be members of any of the societies.

MANUSCRIPT HANDLING

The journals generally handle manuscripts by similar procedures. Four copies of manuscripts, tables, and figures are required by *Agronomy Journal* and *Journal of Agronomic Education* and should be sent to the managing editor at the headquarters office. Three copies are required for *Crop Science* and should also be submitted to the managing editor at the headquarters office. Contributions to the *SSSA Journal* should be sent to the editor-in-chief and those to the *Journal of Environmental Quality* should be sent to the editor at the addresses indicated in the mastheads of these journals. For these two journals, three copies are required.

Receipt of manuscripts is acknowledged and authors are informed as review, approval or release, and publication occur. Correspondence about revisions to the manuscript occurs between associate editors, technical editors or editors, and authors. A manuscript, when approved by the editor, is prepared for printing by the headquarters office.

Papers may be transferred from one society journal to another with permission of the editors and authors involved.

PUBLICATION CHARGES

Publication charges are assessed for regular volunteer papers and notes accepted by society scientific journals. Current charges are published in the journals' masthead. No publication charges are assessed for invited review papers.

Author alterations in galley proofs are charged separately, and an extra charge is assessed if costs exceed the average for photographic reproduction of line drawings or photographs.

PRIOR PUBLICATION

Manuscripts published in the scientific journals must be original reports. They may not have been published previously or simultaneously submitted to another scientific or technical journal. Whether publication in nontechnical outlets constitutes prior publication is decided on a case by case basis. In general, publication in nontechnical media will be considered prior publication only when all of the data and conclusions are included in the nontechnical media.

NOTES

Short papers covering experimental techniques, apparatus, and observations of unique phenomena are published as notes. Notes are not papers that are unacceptable for publication as regular research papers because of content or state of preparation. Review procedures for notes are the same as those for regular articles. A research manuscript that in the opinion of the editor better fits the criteria of a note may be published as a note if the corresponding author agrees. An article submitted as a note is registered with the usual number followed by a capital "N." It may be changed to another classification with the concurrence of the author. The format for notes (two printed pages or less) is usually less formal than that for full length articles.

LETTERS TO THE EDITOR

All of the societies' journals publish letters to the editor. Letters may contain comments on articles appearing in the journals or general discussions about agronomic research and are limited to one printed page. If a letter discusses a published paper, the author of that paper may submit a response to the comments. This response is generally printed along with the letter at the same time. Published letters must be approved by the editor of the journal and may receive a peer review.

EDITOR-IN-CHIEF

Each society has an editor-in-chief, nominated by the president and confirmed by the board of directors for a 3-year term, who may be reappointed for one additional term. This person has overall responsibility for all publications of the respective society and serves on the intersociety editorial policy coordination committee. The committee is chaired by the ASA editor-in-chief.

The editor-in-chief makes recommendations to the president about appointment and reappointment of editors of journals and other publications and serves as a member of all editorial boards and publication committees. This person may be called on to handle special problems through an appeals process and to perform other editorial duties requested by the board of directors. The ASA editor-in-chief also coordinates the policies and activities of *Agronomy Journal, Journal of Environmental Quality, Journal of Agronomic Education,* and *Crops and Soils Magazine.*

OTHER EDITORS

Technical, associate, and consulting editors for each scientific journal are appointed by the president of the respective society for a 3-year term and may be reappointed once.

AGRONOMY JOURNAL

Agronomy Journal, a publication of the American Society of Agronomy, is now published six times a year. The editorial board consists of the editor-in-chief of ASA, an editor, a managing editor, two technical editors in crops, a technical editor in soils, a technical editor in agroclimatology and agronomic modeling, and a number of associate editors who cover subject matter areas of the agronomic, crop, and soil sciences and statistics.

Articles relating to original research in soil-plant relationships; crop science; soil science; biometry; crop, soil, pasture, and range management; crop, forage, and pasture production and utilization; turfgrass; agroclimatology and agronomic modeling are published in *Agronomy Journal* subsequent to review and approval by the editorial board. Articles should make a significant contribution to advancement of knowledge or toward a better understanding of existing agronomic concepts. The study reported must be of potential interest to a significant number of scientists and, if specific to a local situation, must be relevant to a wide body of knowledge in agronomy. Additional details on requirements for articles are published in *Agronomy Journal* each year.

By invitation from the editorial board, review papers may be printed in the journal. Invitational papers from nonmembers may be published on approval by the president if found acceptable by the editorial board.

Responsibilities

1. Editor

The editor, chair of the editorial board, is responsible for overall quality of the content of the journal, and implements policy decisions. The editor and editorial board decide on procedures for manuscript submission, review and referee criteria, acceptance, release, and publication. The editor delegates editorial functions to other members of the editorial board and takes an active part in defining the journal's aims, policies, and editorial coverage. He also handles the appeal procedure for manuscripts that are rejected. The editor may write editorials or solicit manuscripts on special topics.

2. Technical Editor

Technical editors are responsible for the technical and intellectual content of the journal in their assigned areas. They direct the work of assigned associate editors in reviewing and evaluating manuscripts submitted to the journal. They may delegate to associate editors the responsibility of corresponding and working with authors when revisions of manuscripts are needed. Technical editors notify authors when manuscripts are unworthy for publication and inform the managing editor and editor of this action. The managing editor notifies authors of acceptance of their manuscripts for publication.

3. Associate Editor

Associate editors are responsible for obtaining a minimum of two reviews and for evaluating, in a timely manner, the technical and intellectual content and suitability of manuscripts assigned to them. Associate editors make a recommendation to the technical editor about a course of action regarding the disposition of assigned manuscripts but do not approve or reject manuscripts.

4. Managing Editor

The managing editor for *Agronomy Journal* is assigned by the executive vice-president. This person supervises the steps in registering manuscripts, distributing papers to technical editors, copyediting papers for publication, typesetting, sending galley proofs to authors, preparing papers for printing, and producing reprints. The managing editor keeps records of the status of manuscript review.

Manuscript Handling

1. General Procedures

Manuscripts should be submitted in quadruplicate to the managing editor who notifies the corresponding author of receipt of the manuscript and assigns its registration number. The registration number must be used in all correspondence regarding the manuscript. The managing editor assigns the manuscript to a technical editor on the basis of the subject matter. The technical editor, in turn, assigns properly prepared manuscripts to an associate editor. The associate editor obtains a minimum of two reviews.

If the reviewers recommend publication without change and the associate editor agrees, the manuscript and reviewer reports are sent to the technical editor for concurrence.

If the reviewer and the associate editor find that the manuscript could be published after some revision, the manuscript is returned to the author to obtain a satisfactory revision.

The author of an approved manuscript will be notified by the managing editor of the probable publication date. A Permission to Print and Reprint form is sent to the author (see Chapter 9).

If the reviewers and associate editor recommend that a manuscript be rejected, the manuscript and reviewers' comments are sent to the technical editor. If the technical editor concurs that the manuscript should be rejected, the technical editor releases the manuscript to the author.

If a manuscript returned to an author for revision is not returned within a time specified by the journal, it will be released by the technical editor. Once released, manuscripts must be resubmitted to the managing editor at the headquarters office to receive additional consideration by the journal.

2. Preparation of Papers for Publication

Approved manuscripts are prepared for printing in order of their received dates. Manuscripts are read and edited by the managing editor or an assistant editor. The author may be contacted concerning editorial problems.

Galley proofs are sent to the authors along with the manuscript, figure proofs, table proofs, and a reprint order form showing the page and publication charges.

About 15 days are allowed for reading and return of these proofs. Air mail is used to send proofs to foreign countries and should be used for their return.

The production and shipping of reprints is managed by the headquarters office. About 6 weeks are required from delivery of the journal to shipment of reprints.

JOURNAL OF AGRONOMIC EDUCATION

The *Journal of Agronomic Education* (JAE) is published at least once each year by ASA. The journal accepts reports of original studies pertaining to concepts of resident, extension, and industrial education in crop and soil sciences. Authors of manuscripts submitted to JAE are not required to be members of ASA or any of the other societies. Reviews or digests of a comprehensive and well-defined scope are acceptable. The journal also prints notes, articles describing slide sets, book reviews, and letters to the editor. Articles may confirm and strengthen the findings of others, revise established ideas or practices, or challenge accepted theory, providing the evidence presented is significant and convincing. Manuscripts based mainly on personal philosophy or opinion are acceptable if they conform to the above criteria.

The editorial board consists of the editor-in-chief of ASA, an editor, a number of associate editors, and a managing editor.

Manuscripts intended for JAE must be prepared according to the instructions in this manual and sent in quadruplicate to the managing editor at headquarters office. The managing editor notifies the corresponding author of receipt of the manuscript and assigns its registration number. The registration number must be used in all correspondence regarding the manuscript. The managing editor sends the manuscript to an associate editor who obtains a minimum of two reviews. The duties of the editorial board and the remaining steps toward publication are similar as those for *Agronomy Journal*. Appeals of decisions are handled by the editor-in-chief of ASA.

JOURNAL OF ENVIRONMENTAL QUALITY

The *Journal of Environmental Quality* (JEQ) is published quarterly by the ASA, CSSA, and SSSA. The JEQ editorial board consists of the editor-in-chief of ASA, an editor, associate and consulting editors, and a managing editor.

Consulting editors are not required to be ASA members. Their responsibilities and those of the associate editors are similar to those listed for associate editors of *Agronomy Journal*.

Authors of manuscripts submitted to JEQ are not required to be members of the associated societies. Contributions reporting original research or brief reviews and analyses dealing with some aspect of environmental quality in natural and agricultural ecosystems are accepted from all disciplines for consideration by the editorial board. Manuscripts may be volunteered, invited, or coordinated as a symposium. Acceptance of a group of symposium manuscripts for collective publication is limited to one such group per issue. Book reviews may be invited by the editor.

Papers published in JEQ may report the effect of specific practices and substances from agricultural or nonagricultural origin upon environmental quality in natural or agricultural ecosystems, or on the quality of products from these systems; biological methods of pest control; integrated pest control programs; use of soil for the application of agricultural, municipal, or industrial wastes for resource utilization or waste management; and environmental quality aspects of land use and development. Short papers concerning experimental observations or development of methods will be treated as technical reports.

Contributions to JEQ should be submitted in triplicate to the editor. All manuscripts deemed suitable for review are given a registration number that specifies whether the manuscript is a technical report or review and analyses paper.

The editor notifies the corresponding author of receipt of the paper and assigns its registration number. The registration number must be used in all correspondence regarding the manuscript. The editor assigns each manuscript to an associate or consulting editor according to field of specialization.

The remaining procedures are essentially the same as described for *Agronomy Journal*. Appeals of decisions by the editorial board are handled by the editor-in-chief of ASA.

CROP SCIENCE

Crop Science, published bimonthly, is the official publication of the Crop Science Society of America. The publication is prepared by an editorial board consisting of an editor and editor-in-chief, four technical editors, associate editors, and a managing editor. Two technical editors handle manuscripts in crop genetics, breeding, cytology, and statistics. One technical editor handles papers in crop physiology and metabolism, and another technical editor ecology, production, and utilization.

Crop Science publishes original research in crop genetics, breeding, cytology, physiology, management, ecology, quality, utilization, and pest management. Critical reviews may also be published. Cultivar, germplasm, and parental line registrations are published after review by a crop registration committee. Authors should submit registrations to the crop registration subcommittee member for the specific crop or to the crop registration committee chair. All papers, whether invited or volunteered, are subject to review. Additional details on requirements for articles are published in *Crop Science* each year. The rest of the procedures and responsibilities of editorial board members are similar to those of *Agronomy Journal*.

SOIL SCIENCE SOCIETY OF AMERICA JOURNAL

The *Soil Science Society of America Journal*, published bimonthly, is the official publication of the Soil Science Society of America. The editorial board consists of an editor-in-chief, associate editors (including at least one representative for each division of SSSA), and a managing editor.

The *SSSA Journal* publishes papers on original research, reviews of research, and comments and letters to the editor. Papers of appropriate subject matter of two printed pages or less may be submitted as notes. Invitational papers may be published in the journal if accepted by the editorial board.

Contributions to the *SSSA Journal* should be submitted in triplicate to the editor-in-chief. For further details, see SSSA Publication Policy that is published in the first issue of each volume. The editor-in-chief notifies the corresponding author of receipt of the manuscript and assigns its registration number. The registration number must be used in all correspondence regarding the manuscript. The editor assigns each manuscript to an associate editor according to field of specialization. The remaining steps are essentially the same as those described for *Agronomy Journal*, except that associate editors have the authority to approve manuscripts for publication. Manuscripts judged to be unsuitable for publication are referred back to the editor-in-chief. The editor-in-chief handles the appeal procedure for rejected manuscripts.

CHAPTER 3

Procedures for *Crops and Soils Magazine*

Crops and Soils Magazine is a medium of communication for research and field-based agronomists, farmers and their advisors, and persons in agribusiness. The magazine differs from the five scientific journals of the societies in that (i) it is written for a different audience; (ii) most readers are not members of the societies and membership is not required of authors; (iii) no page charge is required in order to publish, and the magazine pays an honorarium to authors of feature and forum articles; and (iv) *Crops and Soils Magazine* often publishes articles outside the scope of the agronomic sciences.

THE EDITORIAL BOARD

The editorial board is composed of a crops and a soils specialist from each U.S. region, plus a chair, the editor, the ASA editor-in-chief, and the directors of advertising and circulation. The chair and board members are appointed to 3-year terms by the ASA president and may be appointed for one additional term.

The board is responsible for the editorial quality of the magazine, working with the editor to keep the magazine up-to-date, and promoting the publication. The chair presides over board meetings and provides advice and counsel on article selection, technical content, and other questions.

TYPES OF ARTICLES

Three major types of articles are published: features, forums, and short articles. Feature and forum articles are written by scientists or specialists. These articles may be volunteered or requested by the editor.

Features report original research or reviews of current research topics, are written by scientists, and are reviewed by one or more members of the magazine's editorial board.

Forum articles may not be reviewed and they may contain opinions on subjects vital to crop growers and agriculturists. They may provoke, stimulate, speculate, and predict.

Short articles are usually written by the magazine staff but may be written by others. No honorarium is paid. Most are based on news releases from universities and industries.

In addition, four other sections regularly appear in the magazine:

1. Events—state, regional, national, and international meetings.
2. The Market—new products and news of mergers, expansions, and other changes within industry.
3. Publications—technical and nontechnical publications, and often a review or listing of a recent book.
4. Abstracts—a brief summary of selected abstracts from current society journals.

ARTICLE FORMAT

The format is that of a magazine, not a technical journal. Informal sections are used. Articles often begin with the results and conclusions of the research and the rest of the information bolsters the conclusions. The language is informal and nontechnical. Documentation is given informally within the text instead of in references at the end. The units of measurement used are those that are most familiar to the reader.

HOW TO SUBMIT MATERIAL

Articles should be submitted to the editor of *Crops and Soils Magazine* at the headquarters office at any time. Before writing articles, authors are encouraged to query the editor by a phone call or letter.

LENGTH OF ARTICLES

A typical feature or forum article is four to six pages of double-spaced typewritten text, plus charts, graphs, and photographs. Short articles normally are two double-spaced typewritten pages in length. Longer articles are generally published as short features. Manuscripts do not need to be on line-numbered paper.

TABLES, ILLUSTRATIONS, AND PHOTOGRAPHS

Tables are used only when necessary. Graphs, charts, and detail drawings are encouraged if they can help present the information.

Use of photographs is encouraged, especially those that show marked contrasts (a fertilized plant beside a check plant; a new piece of equipment beside an outmoded one), a before and after comparison, or a step by step procedure. Photographs should be as interesting as possible. For example, a photo of a pasture is much more interesting if livestock are in it, or a photograph of equipment is more interesting if it shows a person operating it. *Crops and Soils Magazine* can publish only black and white photographs.

OTHER SUBMISSION POINTERS

Articles should be timely. For example, articles on planting crops are published in late winter or spring; articles on crop management in late spring and early summer; articles on crop maturity and harvesting in late summer and fall; and articles on drainage, crop storage, land reshaping, future crop planning, etc. after harvest and into the winter.

A manuscript should be submitted at least 4 months before the desired publication date.

REVIEW, EDITING, AND REWRITING

When a manuscript for a feature article (volunteered or invited) is received, the editor determines whether it is of acceptable subject matter. If it is, a copy is sent to the chair of the editorial board, who obtains technical reviews. Manuscripts are reviewed for technical accuracy by two or more persons with expertise on the subject of the manuscript. The reviewers make their recommendations to the editor.

If the article is approved, the editor begins editing/rewriting. Once finished, a typescript is returned to the author for review, along with comments from the editor and a Transfer of Copyright Form.

CHAPTER 4

Procedures for Monographs, Books, and Other Publications

In addition to the five scientific journals and *Crops and Soils Magazine*, the societies publish monographs, books, special publications, *Agronomy Abstracts*, *Agronomy News*, an annual meeting program, slide sets, glossaries, and miscellaneous papers and booklets. At the time of this writing, CSSA and SSSA are considering initiation of a CSSA and SSSA monograph series. Initially, the same procedures and policies used for agronomy monographs would apply to the CSSA and SSSA monograph series.

MONOGRAPHS

A monograph is a detailed, scholarly treatise written by experts on a single topic or commodity. They are published on an irregular schedule only when a committee of specialists determine a need for monographic treatment of a topic. Plans for monographs are formulated by committees of qualified society members. The final editing, printing, advertising, and distribution of each monograph is handled by the headquarters office.

A monographs committee reviews suggestions for new monographs and revision of existing ones. It is composed of seven members appointed by the president and the editor-in-chief of the society who serves as an ex-officio member of the committee. Members of the monographs committee are appointed for 3-year terms and may be appointed for one additional term. One member of the monographs committee functions as chair for a 1-year term.

When a topic for a new monograph or revision of an existing monograph is proposed, the monographs committee reviews the topic. If the committee recommends that the subject merits further study, it may submit its recommendation to the appropriate executive committee or may suggest the names of scientists for appointment to an ad hoc "monograph feasibility committee." The feasibility committee reports to the monographs committee on (i) need for the proposed monograph; (ii) quantity and quality of research information available on the monograph topic; (iii) existence of, or plans for, books on the same topic; and (iv) potential sales volume.

Based upon the feasibility committee report, the monographs committee may recommend that a monograph on the topic be prepared. The societies' executive committee(s) or board(s) makes the final decision on publication of the monograph.

The monographs committee nominates persons for membership on specific monograph editorial committees to the appropriate society president who appoints an editor and editorial committee to develop the monograph. The monographs committee reviews outlines from the editor.

Duties of the Editor and Editorial Committee

The editor and editorial committee are responsible for the preparation, review, and editing of the monograph. This includes determining the scope of the monograph, organizing subject matter, selecting qualified authors, providing uniformity in style, and technical editing of manuscripts.

The editor and members of the editorial committee may serve as authors, and an author may prepare more than one chapter. The editor of the monograph advises authors on the scope and intended audience for the monograph. Authors do not need to be members of any of the societies. Authors should be apprised of their responsibilities relating to completion of manuscripts and give written consent to prepare the manuscript within a prescribed time.

The editorial committee:

1. Provides authors with instructions on cross referencing, writing style, use of up-to-date common and scientific names, preparation of graphs and figures, and other aspects of manuscript preparation.
2. Sets deadlines for approval of outlines, first drafts, and other stages of preparation.
3. Reviews and edits manuscripts, including making suggestions for coordination, elimination of duplication, and incorporation of topics to make the monograph complete.
4. Submits all manuscripts to headquarters office for style editing.
5. Prepares a preface for the monograph, giving the purpose for publication, intended audience, information on approach in treating subject matter, and acknowledgments.
6. Prepares a subject index for the monograph.

Duties and Responsibilities of Authors of Chapters of Monographs

Authors are responsible for preparing and submitting detailed chapter outlines; first drafts of the manuscript, typed double-spaced on paper with numbered lines; final draft of the manuscript; corrected galley proofs; a subject index, if requested by the editor; and written permission from the owners to use any copyrighted material.

Manuscripts for monograph chapters should be submitted according to deadlines agreed upon with the editor. The editor or editorial committee may replace authors who do not meet deadlines or who provide unsatisfactory manuscripts.

Authors should prepare complete, up-to-date, definitive chapters covering the assigned subject matter. They are responsible for the interpretation they place on the published literature, and should make critical analyses of reported research results. Authors should obtain reviews of their chapters before submitting manuscripts to the editor. The editor should know who reviewed the manuscripts.

Authors are responsible for the costs involved in preparation of their manuscripts, including illustrations. They must agree that material in the manuscript will be published first by the associated societies and that the societies, as publishers, will control its subsequent distribution (see Chapter 9).

Authors should use this manual as the official guide for preparing their manuscripts. When the requirements differ for the various societies, those given for ASA publications should be followed. The editor should inform authors of any special procedures to ensure uniformity in style of writing for the text, units of measurements, scientific names, literature references, illustrations, etc.

Each author receives a complimentary copy of the book. The senior author of a chapter receives 200 reprints of a chapter that he/she shares with coauthors. No honorarium is paid to authors or editors.

Duties of Headquarters Staff

When all manuscripts have been received, the managing editor has responsibility for completing the monograph. The managing editor corresponds directly with the editor and corresponding authors about questions requiring their attention. Galley proofs of each chapter are sent to authors for proofreading. The managing editor forwards page proofs to the monograph editor for completion of the index. The editor prepares a preface and drafts a foreword for approval by the presidents of the societies sponsoring the monograph.

The headquarters office secures a copyright for the book. All promotion, sales, and distribution are the responsibility of the staff at headquarters office.

BOOKS

Subject matter of books is usually not covered in as much depth as in monographs and covers a broader aspect of a particular subject than a special publication.

Books can be a compilation of papers presented at special society(ies)-sponsored symposia or conferences, a compilation of papers written by a number of authors on a particular subject but not presented at a symposium, or a manuscript written by a single author. The societies may consider for publication unsolicited manuscripts that further the educational goals of the societies.

Proposals for books should be submitted to the executive vice president. The executive committees make a decision on publication after consultation with the respective editors-in-chief.

If the papers for the book are presented at a special symposium, a symposium planning committee is appointed by the societies' presidents. The committee develops sessions, outlines topics, and selects participants who become authors of the papers. The executive committee appoints an editorial committee, in consultation with the respective editor-in-chief, which reviews manuscripts. Approval of a publication does not constitute approval of the manuscript(s); after review, manuscripts are approved by the editor of the publication. See also the monographs section for duties of the editor, editorial committee, authors, and headquarters staff.

SPECIAL PUBLICATIONS

Paperbound special publications usually result from timely topics presented from symposia at the annual meetings of the societies. Each society has its own special publication series. The ASA, CSSA, and SSSA can jointly publish any of the series.

Guidelines

1. Organizing Committee

Members who initiate a special publication submit a proposal for a special publication on a form available from headquarters office to the executive vice president. If the proposal is to publish the proceedings of a symposium, it must be submitted at least 6 months before the symposium. Proposals are submitted to the society program officer, who, in turn, submits them to the executive vice president.

The executive committees of the three societies review special publication proposals, and upon consultation with the editors-in-chief of the societies, approve or disapprove publishing the special publication. For a special publication sponsored jointly by all three societies, ASA assumes leadership. If the publication is jointly sponsored by ASA and CSSA, CSSA assumes leadership. If ASA and SSSA are joint sponsors, SSSA assumes leadership.

2. Editorial Committee

The executive committee may appoint an editorial committee for a special publication from persons recommended in the proposal. The editor should keep the editor(s)-in-chief of the societies apprised of the publication's progress. The editor-in-chief of the leadership society serves as an ex-officio member of the editorial committee.

3. Handling of Manuscripts

Papers for a special publication, whether invited or volunteered, must be submitted to the organizer of the symposium or the editor of the special publication at least 4 weeks before the symposium.

Manuscripts submitted to the chair of the editorial committee are assigned to members of the committee for review as outlined below. Manuscripts should meet society publication standards. The chair makes the final decision on the acceptability of manuscripts.

Timeliness of publication of symposia papers is crucial to fulfilling the purposes justifying publication, commitments to authors, and successful sales. Hence, editorial work must be completed promptly and manuscripts sent to the managing editor, normally within 8 weeks after the date of the symposium. Delays greatly exceeding this will cause reevaluation of the publication commitment by the society executive committee.

4. Review of Manuscripts

Manuscripts that are primarily presentations of original data are reviewed to determine that (i) experimental methods and procedures are adequately described, (ii) interpretations and conclusions are valid and consistent with the data presented, and (iii) the paper presents significant new information or significant new interpretation of preexisting data. The authors of such papers must submit a Transfer of Copyright to the society (see Chapter 9).

Contents of manuscripts that are primarily reviews and interpretations of published data are reviewed to determine that (i) all aspects of the topic have been adequately covered, (ii) data presented are representative of the information published, and (iii) the interpretations and conclusions made are consistent with the data and results presented. Authors of such papers must obtain permission to use existing material from the author(s) and publishers.

Reviewers of manuscripts being considered for publication have 4 weeks to review the manuscripts. The authors have 4 weeks to make the revisions and return the manuscripts to the editor of the special publication.

AGRONOMY NEWS

Agronomy News is the official monthly newsletter of the three societies. Members and nonmembers alike may submit items for publication under the following categories:

1. People (names and events in the news)
2. Calendar
3. Meetings
4. Publications
5. Retirements
6. Necrology
7. Personnel (positions available and assistantships)
8. Lists of theses and dissertations (annually)

The newsletter is published monthly. The managing editor should be contacted for specific deadlines. Articles may be submitted by individuals or organizations. Publication charges are not assessed for publishing articles.

Photographs are an excellent addition for many articles. Although color photos are acceptable, black and white photos are preferred. Gray photos reproduce poorly. The background in photos should be simple. Refer to the chapter on illustrations in this manual for further information and to the *Associated Press Stylebook and Libel Manual* (Angione, 1977) or a recent issue of *Agronomy News* for help in writing material for the newsletter.

CHAPTER 5

Preparing the Manuscript

Manuscripts submitted to society journals and other publications are critically reviewed before they are published. The purpose of the review is to assure readers that the papers have been found acceptable by competent, independent professionals. The process often results in desirable changes in the manuscripts, but it is not a substitute for the author's maximum efforts to present the best possible report of findings.

Every paper should have a thorough review by competent colleagues of the author before it is submitted for publication.

Authors should study this section, applicable writing sections in the *CBE Style Manual*, and recent publications of the society before preparing manuscripts. Any paper submitted but not conforming to acceptable standards will be returned to authors for reworking before final review and publication.

The formats used in society journals, books, special publications, and other media differ. This chapter covers formats that can be readily identified and described, but the discussion applies to other formats to varying degrees. The following chapter (Chapter 6) should also be consulted for details of style because the principles of precision, logic, and clarity in presentation expounded here and in that chapter pertain to all manuscripts submitted to the society.

MANUSCRIPT FORMAT FOR JOURNAL ARTICLES

Manuscripts prepared for society journals should usually be arranged in the following order:
1. Title and author(s)
2. Abstract, with additional index words.
3. Introduction (includes literature review). This section is usually not labeled with a section heading.
4. Materials and methods
5. Results. This section is sometimes combined with the discussion.

6. Discussion. This section may be combined with a conclusions section. A separate summary section should not be used because it will duplicate information in the abstract, but a summary statement can be given at the end of the manuscript.
7. References
8. Tables and figures

Other headings may be used, e.g., theory, description of study area, soil profile descriptions, in certain papers. Limited use of subheadings is encouraged to highlight significantly different aspects of the methods, results, or discussion sections. The review articles and notes are usually less formal than full-length articles. Notes must be two printed pages or less. Letters to the editor should not exceed one printed page and are not divided into sections.

Title and Author(s)

The title should represent the article's content and facilitate retrieval in indexes developed by secondary literature services.

A good title (i) briefly identifies the subject, (ii) indicates the purpose of the study, and (iii) gives important, high impact words early. A person usually decides to read an article based on the title's content.

Besides being descriptive, titles should be short. The Societies recommend that titles not exceed 12 words, except in unusual circumstances. A title containing fewer than five words probably should be expanded.

The meaning and order of words in a title are also important. The title must be useful in itself as a label. The terms in the title should be limited to those words that give significant information about the article's content.

Many readers peruse the titles in a table of contents to decide whether or not to turn to a given Abstract. The title must interest these readers. Overly specific, narrow titles with words understandable only to specialists will be passed over. Further, literature searchers will ignore titles that are incomprehensible to all but a few individuals.

Titles should never contain abbreviations, chemical formulas, proprietary names, and jargon. The author should also avoid unusual or outdated terminology in titles.

For economy of space, common names of chemicals and crops should be used in titles. If a crop or microorganism has no common name, then the scientific name (genus and species) is used.

Series titles are used infrequently in society journals. An author contemplating whether or not to prepare a series of articles on the same subject should refer to current editorial policy of the journal. Articles in a series are not discouraged on that basis alone, but the editors need to be assured that all papers in the series are available for review and that the reader can obtain previously and later published material in that series. The series publication of papers also presents scheduling problems in production of the journal.

Authors' names or initials are published as they appear on the manuscript. Footnotes 1 and 2 are always used to document the origin of the paper and to give the professional title(s) of the author(s).

Abstract

A person reading the abstract should be able to tell quickly the value of the report and whether to read further. An abstract is generally read by many more people than those reading the entire report. Thus the abstract has the dual function of supplying information to those who will read the entire report and to those who will read nothing further of the article.

The abstract should be a suitable literary adjunct to the printed article. It should be written after the article is completed and should be consistent with statements in the article. To some extent the abstract will repeat wording in the article, but because it is sometimes read immediately before the introduction or other main sections, it should not be a tedious recapitulation.

On the other hand, the abstract must be completely self-explanatory and intelligible in itself. It should include the following:

1. Reason for doing the work, including rationale or justification for the research.
2. Objectives and topics covered.
3. Brief description of methods used. If the paper deals mainly with methods, give the basic principles, range, and degree of accuracy for new methods.
4. Results.
5. Conclusions.

The abstract also should call attention to new items, observations, and numerical data and should include scientific names for cultivars.

Abstracts should be informative. Expressions such as "is discussed" and "is described" should rarely be included. Specific rather than general statements must be used, especially in the methods and results sections of the abstract. For example, do not say "two rates of P" but say "rates of 40 and 80 kg of P ha^{-1}."

The length of the abstract should not exceed 250 words for full-length papers and 100 words for notes and is not divided into paragraphs. It should not include bibliographic, figure, and table references. Equations, complex equations and formulas, obscure abbreviations, and acronyms also are inappropriate. The scientific names of plants, insects, etc., and full chemical names, must be included in the abstract when the common names are first mentioned.

Additional Index Words

The additional index words are used for two purposes. Journal editors use them in the annual and cumulative indexes. Secondary and indexing services outside the society use them for information storage and retrieval. Together with the title and abstract, the additional index words comprise most of the content of the article that will be available to people looking for it from among thousands of other contributions.

Additional index words should amplify the title, not supplant or repeat it. Scientific names and authorities, when not given in the title, should be given as additional index words. If other plants or organisms important to the subject appear in the paper but not in the title, their scientific names should also be included as additional index words. To choose these words, the author should read through the manuscript for significant key words, noun phrases, and noun clusters that characterize the study. From five to seven additional index words or phrases should be provided. These words should follow the abstract. Phrases over three words should not be used.

Introduction

The article should begin by clearly identifying its subject. The author should state early the hypothesis or definition of the problem the research was designed to solve. A reader needs orientation to the research being reported by brief reference to previous concepts and research. On the other hand, most readers do not need long literature reviews, especially of old references if newer ones are available, or to be convinced about the importance of the research. The purpose of the introduction is to supply sufficient background information to allow the reader to understand and evaluate the results of the present study without needing to refer to previous publications on the topic.

Introductions should be short and include:
1. A brief statement of the problem that justifies doing the work or the hypothesis on which it is based.
2. The findings of others that will be challenged or developed.
3. An explanation of the general approach and objectives. This part may indicate the means by which the question was examined, especially if the methods are new.

References to literature should be limited to information that is essential to the reader's orientation.

Materials and Methods

The purpose of this section is to give enough detail so that a competent worker can repeat the experiments.

For materials, the author should supply the appropriate technical specifications and quantities and source or method of preparation. If necessary, the pertinent chemical and physical properties of the reagents should be listed. Chemical rather than trade names are preferred. Plants, soils, animals, and other organisms should be identified accurately by genus, species, cultivar, series, and special characteristics.

Methods should be cited by a reference if possible. If the techniques are widely familiar, use only their name. If the method is modified, an outline of the modification should be given unless the modification is trivial. Give details of unusual experimental designs or statistical methods. This section may be arranged chronologically, by a succession of techniques, or in another manner.

This section may include tables and figures.

Results

Tables, graphs, and other illustrations in the results section should provide a clear understanding of representative data obtained from the experiments. Data included in illustrations and tables should not be extensively discussed in the text, but significant findings should be noted. When only a few determinations are presented, they should be treated descriptively in the text. Repetitive determinations should be presented in tables or graphs.

A common fault is to repeat in prose what is already clear from a cursory examination of the graphics. If the tables and figures are well constructed, they will show both the results and the experimental design. The objective of each experiment should be made clear in the text. Call attention to special features, such as one quantity being greater than another, that one result is linear over a range, or the optimum value. Finally, the results should be connected to one another. Frequently this causes the results section to be combined with the discussion section.

Discussion

The discussion section interprets data presented in the results section, giving particular attention to the problem, question, or hypothesis presented in the introduction. A good discussion will contain:

1. Principles, relationships, and generalizations that can be supported by the results.
2. Exceptions, lack of correlation, and definition of unsettled points—gap areas or areas needing further investigation.
3. Emphasis on results and conclusions that agree or disagree with other work.
4. Practical as well as theoretical implications.

5. Conclusions, with summary of evidence for each one.

The discussion section, if not combined with the results section, should not recapitulate results, but it should discuss the meaning of the results. The reader should understand how the results provide a solution to the problem stated in the introduction or given as the objective of the work. The work should be connected with previous work and how and why it differs or agrees. References should be limited to those that are most pertinent. Older references should be omitted if they have been superceded by more recent ones.

Speculation is encouraged but should be reasonable, firmly founded in observation, and subject to tests. It must also be identified apart from the discussion and conclusion. Where results differ from previous results for unexplained reasons, possible explanations should not be belabored. Controversial issues should be discussed clearly and fairly.

A common fault of discussion sections is a tendency toward too much contemplation of nonessentials. Only discussion that illuminates significant areas should be presented.

Some papers may warrant a separate conclusions section. It is desirable to present conclusions as part of the discussion section in a paper of average complexity where conclusions are few. Whether this section is combined or separate, the author should include any significant conclusions that have been drawn from the work. These conclusions should be carefully worded so the readers can identify them and will not misunderstand them.

References

The references section lists the literature cited in the paper. Authors are encouraged to cite only published, significant, and up-to-date references in their papers. This section is discussed in more detail later.

Tables and Figures

Supporting data usually are presented in tables and figures. Chapter 7 in this manual contains a discussion of their preparation.

MANUSCRIPTS SUBMITTED FOR OTHER SOCIETY PUBLICATIONS

Journal article format is usually not used in other publications of the societies, but certain sections, such as references, do follow the conventions established for the journal article. Other sections, such as discussion, may be called something else in book chapters, etc., but should adhere to the same scientific and editorial requirements as journal articles.

DETAILS OF MANUSCRIPT PREPARATION

Typing

The manuscript must be typed, double-spaced, on good grade, 216 by 280-mm (8½ by 11 inch) paper or prepared by other techniques to obtain a copy providing comparable quality. Manuscripts produced by dot matrix printers are acceptable if they are of good legible quality. The printer should have true subscripts and superscripts rather than compressed ones. Print should be as dark as that produced by good quality typewriters with dark ribbons. Letters should be completely formed and even on a line to avoid distortion.

Manuscripts must be on line-numbered paper. Line-numbered paper may be purchased from headquarters office or other line numbering can be used, e.g., with a word processor. Submit as many copies of the manuscript as required by the publication. All copies must be legible throughout. Photocopies are acceptable if they are as legible as the original. Authors should keep a file copy of their manuscript. Other guidelines are as follows:

1. Double space all typing, including footnotes, references, data in tables, and captions to figures and tables. Use only one side of the sheet. Indent each paragraph five spaces.
2. When revising a manuscript, type a correction or addition of a few words between lines on the original page. For major revisions, retype the paragraphs or pages involved. Deletions of a word or two may be obliterated or clearly marked on the original pages. If a manuscript is retyped, the original copy must be returned along with the revised one.

Estimating Length of Journal Article

About four manuscript pages of elite typescript (12-pitch face) occupies one printed page. One elite typescript page should contain 25 lines 150 mm (6 inches) wide. If the manuscript is typed with pica (10-pitch face) or some other size of type and another length of line, the proportion will vary accordingly.

The ratio of manuscript to printed page of other publications varies, and editors should be consulted about the format requirements.

Space required for figures can be estimated from the size of the originals and the amount of reduction that will be made in preparing negatives and photostats for printing.

Space required for tables can be estimated from the number of lines of headings, subheadings, and numbers in the tables. Ten lines require about 25 mm (1 inch) of column space. If the lines contain more than about 60 numbers or letters, the table will be set two columns wide.

Headings and Subheadings

Authors should examine samples of the publication for which the manuscript is being prepared. For the journals, the main text headings, such as materials and methods, are typed in capitals in the center of the line. Secondary center headings and side headings are typed in capitals and lower case letters. Run-in headings are typed in the normal paragraph position and underscored for printing in italics. Use of subheadings can help divide papers for guiding readers, but excessive use is distracting. Keep subheadings short.

Captions

Figure captions must be typed together on a separate page. Use "Fig." for abbreviation in captions. Table headings must be typed along with the table and do not require a separate page. Number the figure caption page to follow the reference list.

Footnotes

Authors should avoid use of footnotes, except for footnotes 1 and 2 that are required in society publications. Use the following form for footnotes 1 and 2 of journal papers:

Footnote 1 (the title): The documentation of the paper as to the organization that supported the research (name of institution, city, and state). Received date (date supplied by editor).

Footnote 2 (the authors): Professional titles and addresses of authors including current information if your address or title has changed since the time the paper was written.

Other necessary footnotes may be a government disclaimer in reference to a commercial product or product and trade names.

References

In all society publications, only literature that is available through libraries can be cited. Material not available through libraries, such as personal communications or privileged data should be given in text as parenthetical matter. Include source of data and date (R.D. Jackson, 1982, personal communication). Authors are encouraged to cite only significant, published references. Abstracts, theses or dissertations, and secondary materials should be carefully examined by authors before including them in the reference section. Many of these materials are later published in places that are more easily obtained by readers. If possible, authors should cite the more accessible form of these contributions.

Two methods of giving references in the text are acceptable: the name-year system (e.g., Smith, 1983) and the reference number method (e.g., 3). For two authors, name both of them: Jones and Johnson (1983). With three or more authors, use et al.: Smith et al. (1983) or Smith et al. (9). For two or more articles by the same author(s) in the same year, designate them as follows: Brown (1983a, 1983b) or Smith et al. (1983a, 1983b).

Each reference to a periodical publication must include, in order, the author(s), year of publication, full title of the article, publication in which it appears, and volume and inclusive page numbers. (See examples later in this chapter.)

Reference to a book, bulletin, government document, or conference proceedings must give the author(s), year, title, name of editor(s) if appropriate, edition if other than the first, location and dates (if applicable), publisher, city of publication, and number of the volume (if two or more). If specific pages in a book (not entire chapters) are cited, mention them in the text: Weisman (1983, p. 75).

Publications without consecutive pagination (i.e., each issue within the volume begins with page 1) should include the issue number: 11(2)5-10.

Arrange the list alphabetically by names of the first authors and then by the second and third authors as necessary. Two or more articles by the same author (or authors) are listed chronologically; two or more in the same year are indicated by the letters, a, b, c, etc. All single authored articles of a given individual should precede multiple author articles of which the individual is senior author.

Two sources of error occur in reference citation: inaccurate copying of the bibliographic information and compilation of the reference section after the paper is written. Reviewers and editors cannot be expected to verify the accuracy of the literature citations.

Authors, when copying the publication data from a document, should verify their final product against the document. The reference's author name, title, and other parts should exactly match that shown on the original document. When in doubt, the author should consult a reference librarian for the correct bibliographic citation of difficult material. Readers should be able to obtain cited references by presenting the list to a librarian.

The second type of error occurs when authors (i) either do not include a reference cited in the manuscript or have omitted a reference from the text and have left it in the reference list, or (ii) the names and dates in the reference list do not agree with those in the text. Authors are urged to check the alphabetical reference list against the citations in the body of the manuscript before submitting the manuscript for publication.

Entries with the same senior author (e.g., Shotwell below) should be organized by alphabetizing surnames of succeeding coauthors and then by year, when the name is repeated exactly (see 3 and 4 below).

1. Shotwell, O. L. 1984. _____.
2. Shotwell, O. L., M. L. Goulden, and C. W. Hesseltine. 1982. _____.
3. Shotwell, O. L., C. W. Hesseltine, and M. L. Goulden. 1983. _____.
4. Shotwell, O. L., C. W. Hesseltine, and M. L. Goulden. 1984. _____.
5. Shotwell, O. L., C. W. Hesseltine, E. E. Vandegraft, and M. L. Goulden. 1983. _____.
6. Shotwell, O. L., W. F. Kwolek, M. L. Goulden, L. K. Jackson, and C. W. Hesseltine. 1981. _____.
7. Shotwell, O. L., and D. W. Zwieg. 1984. _____.

Do not capitalize the titles of articles, bulletins, or books except proper names, and the first letter of the first word of the title. (See examples of literature citations.)

Abbreviate names of federal agencies when such abbreviations are permitted in this manual and clearly understood (USDA, ARS, SCS). Do not abbreviate state names except after a geographic term such as Madison, WI; Cook County, IL, and then use zip code abbreviations.

Periodical titles should be abbreviated as given in Chemical Abstracts Service (1984).

Dissertations that are available on microfilm or abstract have a number and publication data that must be given in the reference. If available, please supply the dissertation abstract number or University Microfilm number.

Consult the *CBE Style Manual* for examples of the various types of literature citations. A few of the more common types are shown below.

1. Standard Journal Article

Craig, J.R., and A.G. Wollum II. 1982. Ammonia volatilization and soil nitrogen changes after urea and ammonium nitrate fertilization of *Pinus taeda* L. Soil Sci. Soc. Am. J. 46:409–414.

2. Book

Donahue, R.L., R.W. Miller, and J.C. Shickluna. 1983. Soils: An introduction to soils and plant growth. 5th ed. Prentice-Hall, Englewood Cliffs, NJ.

3. Chapter in Book

Moss, J.P., I.V. Spielman, A.P. Burge, A.K. Singh, and R.W. Gibbons. 1981. Utilization of wild *Arachis* species as a source of *Cercospora* leafspot resistance in groundnut breeding. p. 673–677. *In* G.K. Manna and U. Sinhu (ed.) Perspectives in cytology and genetics, Vol. 3. Hindasia Publishers, Delhi, India.

4. Article With No Identifiable Author (avoid use if possible)

Anonymous. 1984. Computer programs from your radio? Agri-Marketing 22(6):66.

5. General Magazine Article

Davenport, C.H. 1981. Sowing the seeds. Barron's. 2 March, p. 10.

6. Technical Report

U.S. Environmental Protection Agency. 1981. Process design manual for land treatment of municipal wastewater. USEPA Rep. 625/1-77-008 (COE EM1110-1-501). U.S. Government Printing Office, Washington, DC.

7. Conference, Symposium, or Workshop Proceedings

Wittmuss, H.D., G.B. Triplett, and D.W. Greb. 1973. Concepts of conservation tillage systems using surface mulches. p. 5–12. *In* Conservation tillage. Proc. Natl. Conserv. Tillage Conf., Des Moines, IA. 28–30 March 1973. Soil Conservation Society of America. Ankeny, IA.

8. Dissertation

Reeder, J.D. 1981. Nitrogen transformations in revegetated coal spoils. Ph.D. diss. Colorado State Univ., Fort Collins (Diss. Abstr. 81-26447).

9. Transactions

Jansson, S.L. 1967. Soil organic matter and fertility. p. 1–10. *In* G.V. Jacks (ed.) Soil chemistry and fertility. Trans. J. Meet. Comm. 2, 4 Int. Soc. Soil Sci. University Press, Aberdeen, Scotland.

10. Translation

Vigerust, E., and A.R. Selmer-Olsen. 1981. Uptake of heavy metals by some plants from sewage sludge. (In Norwegian.) Fast Avfall. 2:26–29.

11. Advances in Agronomy Series

Savant, N.K., and S.K. DeDatta. 1982. Nitrogen transformations in wetland rice soils. Adv. Agron. 35:241–302.

12. Patent

Titcomb, S.T., and A.A. Juers. 1976. Reduced calorie bread and method of making same. U.S. Patent 3 979 523. Date issued: 7 September.

13. Miscellaneous Publication

Wadleigh, C.H. 1968. Wastes in relation to agriculture and forestry. USDA Misc. Pub. 1065. U.S. Government Printing Office, Washington, DC.

14. Corporate Author of Article

American Public Health Association. 1980. Standard methods for the examination of wastewater. 15th ed. American Public Health Association, New York.

15. State Publication

DePuit, E.J., C.J. Skilbred, and J.G. Coenenberg. 1980. Vegetation characteristics on sodic mine spoils. p. 48–74. In D.J. Dollhopf et al. (ed.) Chemical amendments and irrigation effects on sodium migration and vegetation characteristics on sodic mine soils in Montana. Mont. Agric. Exp. Stn. Bull. 736.

16. Federal Publication

Soil Survey Staff. 1975. Soil taxonomy: A basic system of soil classification for making and interpreting soil surveys. USDA-SCS Agric. Handb. 436. U.S. Government Printing Office, Washington, DC.

17. Society Publications

Journals

Sims, J.L., M. Casey, and K.L. Wells. 1984. Fertilizer placement effects on growth, yield, and chemical composition of burley tobacco. Agron. J. 76:183–188.

Monographs

Fryxell, P.A. 1984. Taxonomy and germplasm resources. *In* R.J. Kohel and C.F. Lewis (ed.) Cotton. Agronomy 24:27–57.

Books

Tanner, C.B., and T.R. Sinclair. 1983. Efficient water use in crop production: research or re-search? p.1–27. *In* H.M. Taylor et al. (ed.) Limitations to efficient water use in crop production. American Society of Agronomy, Crop Science Society of America, and Soil Science Society of America, Madison, WI.

Special Publication

Whisler, F.D., J.R. Lambert, and J.A. Landivar. 1982. Predicting tillage effects on cotton growth and yield. p. 179–198. *In* P.W. Unger and D.M. Van Doren (ed.) Predicting tillage effects on soil physical properties and processes. Spec. Pub. 44. American Society of Agronomy, Madison, WI.

Agronomy Abstracts

Knievel, D.P., and C.A. Jones. 1983. Spikelet initiation and abortion and kernel growth rate in maize under nitrogen stress. Agron. Abstr. American Society of Agronomy, Madison, WI. p. 94.

Crops and Soils Magazine

Mulvaney, D.L., and L. Paul. 1984. Rotating crops and tillage. Crops Soils 36(7):18–19.

CHAPTER 6

Conventions and Style

Authors are responsible for making papers clear, concise, and accurate. Manuscripts should be thoroughly reviewed by the author's colleagues before submission to the society. Authors should consult this manual, other society guidelines and instructions mentioned here, and general style manuals when preparing material. Manuscripts that are not properly prepared are returned to authors for correction.

Helpful sources for authors of journal articles, monographs, books, special publications, and other publications of the societies are the *CBE Style Manual*, *The Chicago Manual of Style*, *U.S. Government Printing Office Style Manual* (U.S. Government Printing Office, 1973), and other widely available publications. These books contain details of grammar, punctuation, tables, and other style matters. Authors are also encouraged to study recent issues of society journals and books for the general style and format used.

This manual should be used as a primary source for conventions and style. Other books, such as the ones listed above, supplement this manual.

NOMENCLATURE AND TERMINOLOGY

Biology

The common name, Latin binomial or trinomial (in italics), and the authorship should be shown for plants, insects, animals, and pathogens when first used in the abstract and in the text. Common names, if they exist, should be used in titles without the scientific names.

Scientific names should be in accord with published authorities. For cultivated plants, the rules of nomenclature have been established in the *International Code of Nomenclature for Cultivated Plants* (Brickell, 1980). Various publications, e.g., *Hortus III*, record these names, but revisions are made from time to time because of new taxonomic and nomenclatural evidence. *Crop Science* publishes articles on registered cultivars, germplasms, and parental lines. Crop cultivars (not experimental lines and strains) must be identified by single quotation marks when first mentioned in the abstract and text, e.g., 'Vernal' alfalfa (*Medicago sativa* L.) or *Medicago sativa* L. 'Vernal'. All authorities, including secondary ones, should be cited; e.g., *Glycine max* (L.) Merr. Do not use the word cultivar and single quotation marks at the same time. The abbreviation cv. for cultivars is allowed in the societies' publications. The abbreviation var. (for varieties) refers to botanical varieties and is not appropriate for cultivars. The terms cultivar and variety are synonymous for cultivated plants, but the term cultivar is preferred.

The Crop Science Society of America, through its committee on crop registration, publishes lists of registered field crop cultivars. The most recent is *Registered Field Crop Varieties: 1926–1981* (CSSA, 1982). Copies of this and subsequent lists are available from the headquarters office.

Chemistry

Chemical symbols, including ionic charge, should be used instead of words for elements, ions, or compounds, if possible, e.g., NO_3^- leaching, instead of nitrate leaching, in the text. Use the abbreviation for elements unless they begin a sentence. The ionic charge should be included with all anions and cations, including hyphenated expressions. The ionic charge should be written as superscripts with the value preceding the sign, e.g., Al^{3+}, K^+, Cl^-, NH_4^+-N.

The amounts and proportions of nutrient elements must be expressed in terms of the elements or in other ways as needed for theoretical purposes. The amounts or proportions of the oxide forms (P_2O_5, K_2O, etc.) may be included in parentheses.

Full chemical names for compounds must be used when they are first mentioned in the abstract and in the text. Use the most up-to-date chemical names available. The names can be carried in tabular form if there are a large number. Thereafter, the common or generic name can be used, e.g., atrazine, 2,4-D, etc. Trade names should be avoided whenever possible. If it is necessary to use a trade name, it should be capitalized and spelled out as specified by the trademark owner.

In the USA and Canada the authority for names of chemical compounds is *Chemical Abstracts* and its indexes. The American Chemical Society's *Handbook for Authors* and the *CBE Style Manual* contain many additional details on nomenclature in chemistry and biochemistry. Publications of the American Chemical Society's committee on nomenclature and the nomenclature commissions of the International Union of Pure and Applied Chemistry are available through Chemical Abstracts Service, Columbus, OH.

The section on SI units in this chapter has further information regarding units and concentration.

Information on herbicides is found in the list of "Common and Chemical Names of Herbicides" included on the back covers of the journal *Weed Science* and in the *Herbicide Handbook* published by the Weed Science Society of America (WSSA, 1983).

Many of the organic substances used for pesticides contain prefixes, letters, and numbers designating configuration and rotation of the chemical structure. Some general rules are:

1. The most common hyphenated prefixes that should be italicized (underlined) are *o-, m-, p-, s-, cis-, trans-, sec-, tert-, endo-,* and *exo-*. The prefixes bis- and tris- are not italicized. Elements that occur as locants are also italicized: *O-, S-, N-, H-,* etc.
2. Configurational relationships may be indicated by the italic capital letter prefix *R* and *S*. Rotation may be shown with small capitals D and L, and in some cases the italic *d* and *l* are used.

SPECIALIZED TERMINOLOGY

Professional societies in special fields and the *CBE Style Manual* list terms used in various disciplines. Authors using specialized vocabulary not in published material by those societies or in a good dictionary should consult those societies before establishing new terminology. Committees of ASA, CSSA, and SSSA have studied terminology in specialized fields and have indicated a preference in many cases.

Crop Terminology

A summary of terms compiled by the committee on crop terminology is published at intervals in *Crop Science*. Two reports (Leonard et al., 1968; Shibles, 1976) have been published at this writing. Their reports cite the relevant articles and lists published in related fields and includes the previously published reports issued as mimeographs or articles on definitions by the earlier committees. In addition, letters in the journal comment on various aspects of terminology (e.g., Dybing, 1977).

Identification of Soil

Ideally, both series and family names should be given for soils discussed in the societies' publications. If the series name is not known, give the family name. If the family name is not known, give the subgroup or higher level name. Subgroup names should be singular if reference is to a single pedon or as an entity; otherwise the plural form should be used. Examples are as follows:

The soil used in this study was collected from the A horizon of a Brookston pedon (fine-loamy, mixed, mesic Typic Argiaquoll).

Criteria of the Typic Hapludult subgroup were examined.

Soils of the Ramona (fine-loamy, mixed, thermic Typic Haploxeralfs) and the Kimberlina (coarse-loamy, mixed [calcareous], thermic Typic Torriorthents) series were treated.

All soils used in the experiments were Typic Haplorthods.

For field experiments, soils present in plots or fields should be described including their taxonomic placement. It may also be appropriate to name and briefly discuss the mapping units present in the study areas. Dissimilar inclusions that are present should also be described. An example would be:

> The 5-ha study area is mapped as a Yolo silt loam, 0 to 2% slopes. The Yolo series is a member of the fine-silty, mixed, nonacid, thermic Typic Xerorthents. Dissimilar inclusions of Cortina very gravelly sandy loam (loamy-skeletal, mixed, nonacid, thermic Typic Xerofluvents) are present in small areas.

If possible, members of the National Cooperative Soil Survey (NCSS) should be consulted for proper identification.

Contributors outside the United States are encouraged to give soil identification according to the U.S. system of soil taxonomy (Soil Survey Staff, 1975) in addition to the identification in their national system. (See also Soil Survey Staff, 1978.) The USDA-SCS soil horizon nomenclature should be used to describe soil pedons. A brief discussion and lists of new and old designations for soil horizons and layers was published in the *SSSA Journal* (Guthrie and Witty, 1982).

Light Measurements and Photosynthesis

The term "light intensity" was abandoned to denote the amount of light incident on a surface (Dybing, 1977). The *Crop Science* editorial board also discontinued the use of the photometric system and units scaled to the response of the human eye. Society publications use the radiometric system with SI units denoting the energy or the quantum content of the radiation used by plants (see p. 47).

Terms recommended by the Committee on Crop Terminology for the expression of photosynthetic energy and photosynthetic capacity were defined by Shibles (1976). These are photosynthetically active radiation, photosynthetic photon flux density, photosynthetic irradiance, apparent photosynthesis, and CO_2 exchange rate.

Soil Science Terminology

The *Glossary of Soil Science Terms* (revised SSSA, 1984) contains definitions of 1200 terms, plus appendices covering obsolete terms, procedural guide for tillage terminology, and new designations for soil horizons and layers.

STATISTICAL ANALYSIS AND EXPERIMENTAL DESIGN

Readers of scientific publications must understand how the authors designed and conducted their experiments so that the results can be judged for validity and so that previous experiments may serve as a basis for the design of future experiments. Research design consists of two components: treatment design and experimental or environmental design. Both of these components must be explicitly described when reporting results of experiments in the societies' publications.

Treatment design is used in a broad sense and includes levels of factors in factorial experiments, populations of genotypes used in genetic and physiological experiments, soil types, and many other physical and biological variables. Experimental design refers to the method of arranging the experimental units and the method of assigning treatments to the units. Included should be the number of replicates, description of conditions at field sites and in greenhouse or controlled environment chambers, number of sites and years, and how measurements were made for specific traits. Blocking or other restrictions used in assigning treatments to experimental units should be clearly described. The number of experimental units used and the number of samples taken from each unit should also be clear to the reader.

Special care should be taken in reporting results of field research. First, the results must be obtained using adequate treatment and experimental design, and, as indicated previously, the treatment and design should be clearly described. Second, since field conditions are not repeatable from year to year or site to site, an adequate sample of environments must be used. Usually studies on crop characteristics that are sensitive to environmental effects must be repeated at other site(s) or environment(s) (years) to establish the validity of results. Depending on the nature of the study, two or more environmental regimes may be necessary to obtain meaningful results.

The experimental and treatment designs dictate the proper method of statistical analysis and the basis for assessing the precision of treatment means.

A measure of the precision achieved should be reported for all data on which conclusions are drawn. Two methods for doing this include reporting the standard error of a treatment mean or the coefficient of variation. It is not usually necessary to report standard deviations for individual portions of the total experiment.

When treatments have a logical structure, orthogonal contrasts among treatment means should be made. All possible comparisons among treatment means can be done, but authors should be aware of the limitations of this approach when little information exists on the structure of the treatment (Carmer and Walker, 1982; Chew, 1980; Little, 1978; Nelson and Rawlings, 1983; and Peterson, 1977).

When common experimental designs such as the completely randomized, randomized block, or split-plot designs are used, it is not necessary to cite a reference, but the author should identify the design. It is appropriate, however, to cite references to little-used methods, designs, or statistical analyses. Also, if computer programs are used that are not commonly used, the documentation reference should be cited.

Brief analysis of variance tables with mean squares and degrees of freedom may be published in instances where they are needed for clarity in reporting results. For factorial experiments or for treatments where treatment structure suggests logical contrasts, these tables are an efficient way to summarize the relative importance of the various effects, especially if imbalance in the design was caused by unequal replication. Analyses of variance tables are also useful in reporting mean squares in situations where variance component estimation is the principle objective. In such instances, factors considered to be fixed and those considered to be random should be designated.

Some widely used statistical abbreviations and symbols are given in Table 1. Additional guidelines may be published by the various journals.

Table 1. Some abbreviations widely used in statistics.†

Statistic	Sample Preferred symbol	Acceptable	Population
Arithmetic mean	$\bar{x}$		μ
Chi-square	χ^2		
Correlation coefficient	r		
Coefficient of multiple determination	R^2		
Coefficient of simple determination	r^2		
Coefficient of variation	CV		
Degrees of freedom	df	DF	
Least significant difference	LSD		
Multiple correlation coefficient	R		
Not significant	NS		
Probability of type I error	α		
Probability of type II error	β		
Regression coefficient	b		β
Sample size	n		N
Standard error of mean	SE	$s_{\bar{x}}$	$\sigma_{\bar{x}}$
Standard deviation of sample	SD	s	σ
Student's t	t		
Variance	s^2		σ^2
Variance ratio	F		

† *,** are used to show significance at the $P = 0.05$ and 0.01, respectively. Significance at other levels should be designated by a supplemental note.

MEASUREMENTS, SI SYSTEM

The SI system (Le Système International d'Unités) of reporting measurements is required in all society publications except *Crops and Soils Magazine* and *Agronomy News*. Other units may be reported in parentheses at the option of the author, if this inclusion will clarify interpretation of the data.

Basic references on the use of SI units were published by the National Bureau of Standards and the American Society for Testing Materials. The 1982 issue of *Journal of Agronomic Education* contains three articles on use of SI units and conversions from other units that are particularly useful to agronomists, crop scientists, and soil scientists.

Base, Supplementary, and Derived Units

The SI system is based on seven base and two supplementary units that are listed in Table 2 with their names and symbols.

The definition of the mole (mol), adopted by the 14th General Conference on Weights and Measures in 1971 is the amount of substance of a system that contains as many elementary elements as there are atoms in 12 g of ^{12}C. When the unit mole is used, the entities must be specified.

Derived units (Table 3) are expressed algebraically in terms of base units. Some of these units have been given special names and symbols, which may be used to express still other derived units. An example of a derived unit with a special name is the newton (N) for force. The newton is expressed in basic units as meter kilogram per second squared. Another unit with a special name is the pascal (Pa), which is a newton per square meter.

The SI base unit for thermodynamic temperature is kelvin. Because of its wide use, the Celsius scale may also be used to express temperature. The degree sign should be used with Celsius temperature (°C) but not with the kelvin (K) scale.

Table 2. Base and supplementary SI units.

Quantity	Unit	Symbol
Amount of substance	mole	mol
Electric current	ampere	A
Length	meter	m
Luminous intensity	candela	cd
Mass	kilogram	kg
Thermodynamic temperature	kelvin	K
Time	second	s
Plane angle	radian	rad
Solid angle	steradian	sr

Table 3. Derived SI units with special names.

Quantity	Name	Symbol	Expression in terms of other units	Expression in terms of SI base units
Absorbed dose, specific energy imparted, kerma, absorbed dose index	gray	Gy	J/kg	$m^2\,s^{-2}$
Activity (of a radionuclide)	becquerel	Bq		s^{-1}
Capacitance	farad	F	C/V	$m^{-2}\,kg^{-1}\,s^4\,A^2$
Celsius temperature	degree Celsius	°C		K
Conductance	siemens	S	A/V	$m^{-2}\,kg^{-1}\,s^3\,A^2$
Electric potential, potential difference, electromotive force	volt	V	W/A	$m^2\,kg\,s^{-3}\,A^{-1}$
Electric resistance	ohm	Ω	V/A	$m^2\,kg\,s^{-3}\,A^{-2}$
Energy, work, quantity of heat	joule	J	Nm	$m^2\,kg\,s^{-2}$
Force	newton	N		$m\,kg\,s^{-2}$
Frequency	hertz	Hz		s^{-1}
Power, radiant flux	watt	W	J/s	$m^2\,kg\,s^{-3}$
Pressure, stress	pascal	Pa	N/m^2	$m^{-1}\,kg\,s^{-2}$
Quantity of electricity, electric charge	coulomb	C		s A

Table 4. SI prefixes.

Order of magnitude	Prefix	Symbol
10^{18}	exa	E
10^{15}	peta	P
10^{12}	tera	T
10^{9}	giga	G
10^{6}	mega	M
10^{3}	kilo	k
10^{2}	hecto†	h
10^{1}	deca†	da
10^{-1}	deci†	d
10^{-2}	centi†	c
10^{-3}	milli	m
10^{-6}	micro	μ
10^{-9}	nano	n
10^{-12}	pico	p
10^{-15}	femto	τ
10^{-18}	atto	a

† To be avoided when possible.

Using SI Units

Prefixes and symbols listed in Table 4 are used to indicate orders of magnitude in SI units. They reduce the use of nonsignificant digits and decimals and provide a convenient substitute for writing powers of 10 as generally preferred in computations. With a few exceptions (see Use of Non-SI Units section) base units are required in the denominator of combinations of units, while appropriate prefixes for multiples (or submultiples) are selected for the numerator so that the numerical value of the term lies between 0.1 and 1000. The same unit, multiple, or submultiple should be used throughout the text, tables, and graphs. An exponent attached to a symbol containing a prefix indicates that the unit with its prefix is raised to the power expressed by the exponent, e.g.,

$$1 \text{ mm}^3 = (10^{-3} \text{ m})^3 = 10^{-9} \text{ m}^3.$$

Punctuation is used sparingly with SI units. The center dot, generally used to indicate the product of two or more units, is omitted when there is no risk of confusion with another unit symbol (use N m not N·m). A solidus (oblique stroke,/), horizontal line, or negative powers may be used to express a derived unit formed from two others by division, e.g.,

$$\text{m/s or m s}^{-1}.$$

Only one solidus may be used in combinations of units, unless parentheses are used to avoid ambiguity, e.g.,

$$\text{g m}^{-2} \text{ s}^{-1} \text{ or g/(m}^2 \text{ s) but not g/m}^2\text{/s.}$$

Periods are not used after any SI unit symbol except at the end of a sentence. When numbers are less than one, a zero should be written before the decimal marker (e.g., 0.7).

Use of Non-SI Units

Some units not in SI can be used—including use in the denominator—in the societies' publications, but these units have been limited to those that are convenient for crop and soil scientists. The quantity of area can be expressed as hectare (1 ha = 10^4 m²). Use of liter (10^{-3} m³) in the denominator of derived units is permitted, but m³ is encouraged.

The base unit, second(s), is the preferred unit of time. Other units—minutes (60 s), hour (3600 s), day (86 400 s)—are acceptable although their use introduces difficulties in rapid conversion from one time scale to another. Units of time that vary in length, e.g., month or growing season, should not be used. In expressions of combinations of units, second is required for events that are measured during periods of less than 1 day. For events with measurement periods longer than 1 day, combinations of units may be expressed in terms of days. For example, crop growth rate can be expressed as g m^{-2} day^{-1}.

In SI, a tonne, t, equals 10^3 kg, or 1 Mg, and is understood to be metric ton. Do not use the term "metric ton." Although it is obvious that when SI units are used, tonne cannot refer to the English long or short ton, potential confusion can be avoided by using the unit megagram, Mg. Teragram, Tg, should be used in place of million tonnes where applicable.

Radians is the base unit for measurement of plane angles, but degrees are also acceptable.

Specific Applications

Special attention is required for reporting concentration, exchange composition and capacity, energy of soil water (or water potential), and light. Table 5 summarizes the appropriate units for society publications. Prefixes, other than those shown in Table 5, may be used as noted in Table 4, so that numerical values are between 0.1 and 1000.

1. Concentration

Normality, N, the amount of substance concentration based on the concept of equivalent concentration, should not be used. Concentrations should be expressed on a molar basis as shown below. Examples for correctly expressing concentration (conc) include:

$$\text{conc(HCl)} = 0.1 \text{ mol L}^{-1} = 0.1 \text{ M HCl and}$$

$$\text{conc(H}_2\text{PO}_4^-) = 2.1 \text{ mol m}^{-3} = 2.1 \text{ mmol L}^{-1} = 2.1 \text{ mM H}_2\text{PO}_4^-$$

The concentration 0.1 mol L^{-1} can also be reported as a 0.1 M (molar) solution. Solutions containing ions of mixed valence also should be given on a molar basis of each ion. Molality (mol kg^{-1}) is an acceptable term and unit; it is the preferred unit for precise, nonisothermal conditions.

Gas concentration can be expressed as mol m^{-3}, g m^{-3}, partial pressure, or mole fraction. The denominator of the mole fraction needs no summation sign because SI defines a mole as Avogadro's number of any defined substance, including a mixture such as air. An O$_2$ concentration of 210 mL L^{-1} is therefore 21 × 10^{-2} mol mol^{-1} or 0.21 mol fraction. A CO$_2$ concentration of 335 μmol mol^{-1} equals 335 μmol fraction.

Nutrient concentration in plants, soil, or fertilizer can be expressed on the basis of mass as well as the amount of substance. For example, plant phosphorus level could be reported as 180 mmol P kg^{-1}, or 5.58 g P kg^{-1}.

Table 5. Preferred (P) and acceptable (A) units for several quantities.

Quantity	Application	Unit	Symbol
Concentration	Known molar mass (liquid and solid material)	mole per cubic meter (P)	mol m^{-3}
		mole per kilogram (P)	mol kg^{-1}
		mole per liter (A)	mol L^{-1}
		gram per liter (A)	g L^{-1}
	Unknown molar mass (liquid and solid material)	gram per kilogram (A)	g kg^{-1}
		gram per cubic meter (P)	g m^{-3}
		gram per liter (A)	g L^{-1}
	Known ionic charge	mole charge per cubic meter (P)	mol (+) m^{-3} or mol (−) m^{-3}
		mole charge per liter (A)	mol (+) L^{-1} or mol (−) L^{-1}
	Gas	mole per cubic meter (P)	mol m^{-3}
		gram per cubic meter (A)	g m^{-3}
		gram per liter (A)	g L^{-1}
		liter per liter (A)	L L^{-1}
		microliter per liter (A)	μL L^{-1}
		mole per liter (A)	mol L^{-1}
		mole fraction (A)	mol mol^{-1}
Exchange parameters	Exchange capacity	mole charge of saturating ion, i, per kilogram (P)	mol (i) kg^{-1}
		centimole charge of saturating ion, i, per kilogram (A)	cmol (i) kg^{-1}
	Exchangeable ion composition	mole charge of specified ion, i, per kilogram	mol (i) kg^{-1}
	Sum of exchangeable ions	mole of ion charge per kilogram	mol (+) kg^{-1}
Light	Irradiance	watt per square meter	W m^{-2}
	Photon flux density (400–700 nm)	micromole per square meter second	μmol m^{-2} s^{-1}
Water potential	Driving force for flow	joule per kilogram (P)	J kg^{-1}
		kilopascal (A)	kPa
		meter of water in a gravitational field (A)	m

Water content of plant tissue or plant parts can be expressed in terms of water mass per unit mass of plant material, e.g., g H$_2$0 kg^{-1}. Authors should state whether reported plant mass is on a dry or wet basis.

2. Exchange Composition and Capacity

Historically, soil scientists have expressed exchange capacity in milliequivalents (meq) per 100 g. The units in neither the numerator nor denominator conform to SI. Exchange capacity and exchangeable ion composition should be expressed as moles of charge, either positive (+) or negative (−), per unit mass. For example, the exchangeable ion composition and cation exchange capacity for a soil that contains 4 cmol of exchangeable K$^+$ and 2 cmol of exchangeable Ca^{2+} (4 cmol of ½ Ca^{2+} or 4 cmol of +) per kilogram can be reported in tabular form as follows:

Exchangeable ion		Cation exchange capacity
K$^+$	Ca^{2+}	
-------	------- cmol(+)kg^{-1}	-------
4	4	8

Formerly, the method of expressing exchange composition and capacity would indicate the soil contained 4 meq per 100 g of exchangeable K$^+$ and of Ca^{2+} with cation exchange capacity of 8 meq per 100 g.

If the cation exchange capacity is determined by the single ion saturation technique, the ion used should be specified since it can affect the cation exchange capacity measured. If Mg^{2+} were used for the soil in the example above and specific ion effects were nonsignificant, the cation exchange capacity would be expressed as 8 cmol (½Mg^{2+}) kg^{-1}.

3. Energy of Soil Water or Water Potential

Soil water potential refers to its equivalent potential energy; it can be expressed on either a mass or a volume basis. Energy per unit mass has units of joules per kilogram (J kg^{-1}) in SI. Energy per unit volume is dimensionally equivalent to pressure, and the SI pressure unit is the pascal (Pa). One joule per kilogram is 1 kPa if the density of water is 1 Mg m^{-3}; and since 1 bar = 100 kPa, 1 J kg^{-1} is equal to 0.01 bar at this same density. Energy per unit mass (J kg^{-1}) is preferred over the pressure unit (Pa).

The height of a water column in the earth's gravitational field, energy per unit of weight, can be used as an index of water potential or energy. The potential in joules per kilogram is the gravitational constant multiplied by the height of the water column. Since the gravitational constant is essentially 10 (9.81 m s^{-2}), hydraulic head in meters of water is approximately 10 times the water potential expressed in joules per kilogram or kPa.

4. Light

Accepted SI notation for total radiant energy per unit area is joule per square meter. Energy per unit time or irradiance is expressed in watts per square meter. Alternative units, based on calories or ergs for energy and square centimeter for area, are not acceptable. Also, photometric units, e.g., lux, are not acceptable.

Plant scientists studying photochemically triggered responses, e.g., photosynthesis, photomorphogenesis, and phototropism, may quantify radiation in terms of number of photons rather than energy content. Photon flux density per unit area should be expressed in moles of photons per square meter second. Historically, one mole (1 mol) of photons has been equated with one einstein (1 E). Thus, $\mu E\ m^{-2}\ s^{-1}$ is in common use as a unit of photon flux density but is not acceptable in SI units. The SI units μmol photon $m^{-2}\ s^{-1}$ is equivalent and should be used. The photosynthetic photon flux density (PPFD) is photon flux density in the waveband 400 to 700 nm. For studies involving other wavebands, the waveband should be specified.

5. Use of Percentage in the SI System

Whenever the composition of some mixture is being described and it is possible to express elements of the mixture in SI base or derived units, then use of percent is unacceptable and should be replaced by appropriate SI units. For example, plant nutrient concentration must be expressed in SI units based on either amount of substance or mass.

When the elements of an event cannot be described in SI base or derived units, or when a well-known fractional comparison of an event is being described, percentage is acceptable. In general, percentage can be used to express the fraction of a whole value, when that value is given in SI units. The following are examples where use of percentage is acceptable:

1. Coefficient of variation.
2. Botanical composition, plant stand, and cover estimates.
3. Percent leaves (or plants) infected.
4. Percent increase (decrease) in yield.
5. Percent recovery of applied element(s) by plants, extractants, etc.
6. Fertilizer grades.
7. Percent relative humidity.
8. As an alternate unit for soil texture. This is allowed because each component is well defined and is a fraction on a mass basis.
9. As an alternate unit to express fractional base saturation. This is permissible because each component is a fraction on a chemical basis.

6. Parts per Million

Parts per million is an ambiguous term. To avoid ambiguity, authors are required to use preferred or acceptable units as discussed. Depending upon the type of data, authors could use $\mu L\ L^{-1}$, $mg\ L^{-1}$, or $mg\ kg^{-1}$ in place of parts per million.

7. Cotton Fiber

Official standards for cotton staple length are given in terms of inches and fractions of an inch, generally in gradations of 1/32 of an inch. Stapling is done by a classer in comparison with staple standards. Measurement by instrument has shown unequal increments between consecutive staples in these standards. Because the classer is the authority on length, these unequal increments have been maintained. When staple length is determined by a classer, it may be reported as a code number with the code being the number of thirty-seconds of an inch called by the classer.

Instrument measurements are preferable in experimental work because of equal incremental differences between successive fiber lengths. These values should be reported using appropriate SI units (Table 6). Fiber fineness determined by the micronaire instrument should be reported as micronaire reading.

Recommended Units and Conversion Factors

Tables of recommended units (Table 6) and conversion factors (Table 7) are included to aid in the use of SI units.

Table 6. Examples of preferred (P) and alternate (A) units for general use.

Quantity	Application	Unit	Symbol
Area	Land area	square meter (P)	m^2
		hectare (A)	ha
	Leaf area	square meter	m^2
	Specific surface area of soil	square meter per kilogram	$m^2\ kg^{-1}$
Density	Soil bulk density	megagram per cubic meter	$Mg\ m^{-3}$
Electrical conductivity†	Salt tolerance	siemens per meter	$S\ m^{-1}$
Elongation rate	Plant	millimeter per second (P)	$mm\ s^{-1}$
		millimeter per day (A)	$mm\ day^{-1}$
Ethylene production	N_2-fixing activity	nanomole per plant second	$nmol\ plant^{-1}\ s^{-1}$
Extractable ions	Soil	milligram per kilogram	$mg\ kg^{-1}$
Fertilizer rates	Soil	grams per square meter (P)	$g\ m^{-2}$
		kilogram per hectare (A)	$kg\ ha^{-1}$
Fiber strength	Cotton fibers	kilonewton meter per kilogram	$kN\ m\ kg^{-1}$

† The term, "electrolytic conductivity", has been substituted for "electrical conductivity" by the International Union of Pure and Applied Chemistry (IUPAC). Use of the SI term, "electrolytic conductivity" is permissible but not mandatory in ASA publications at this time. (table cont. on next page)

Table 6. Continued.

Flux density	Heat flow	watts per square meter	W m^{-2}
	Gas diffusion	mole per square meter second (P)	mol m^{-2} s^{-1}
		gram per square meter second (A)	g m^{-2} s^{-1}
	Water flow	kilogram per square meter second (P)	kg m^{-2} s^{-1}
		cubic meter per square meter second (A)	m^3 m^{-2} s^{-1} or m s^{-1}
Gas diffusivity	Gas diffusion	square meter per second	m^2 s^{-1}
Grain test weight	Grain	kilogram per cubic meter	kg m^{-3}
Hydraulic conductivity	Water flow	kilogram second per cubic meter (P)	kg s m^{-3}
		cubic meter second per kilogram (A)	m^3 s kg^{-1}
		meter per second (A)	m s^{-1}
Ion transport	Ion uptake	mole per kilogram (of dry plant tissue) second	mol kg^{-1} s^{-1}
		mole of charge per kilogram (of dry plant tissue) second	mol (+) kg^{-1} s^{-1} or mol (−) kg^{-1} s^{-1}
Leaf area ratio	Plant	square meter per kilogram	m^2 kg^{-1}
Length	Soil depth	meter	m
Magnetic flux density	Electronic spin resonance (ESR)	tesla	T
Nutrient concentration	Plant	millimole per kilogram (P)	mmol kg^{-1}
		gram per kilogram (A)	g kg^{-1}
Photosynthetic rate	CO_2 amount of substance flux density (P)	micromole per square meter second (P)	μmol m^{-2} s^{-1}
	CO_2 mass flux density (A)	milligram per square meter second	mg m^{-2} s^{-1}
Plant growth rate		gram per square meter day	g m^{-2} day^{-1}
Resistance	Stomatal	second per meter	s m^{-1}
Soil texture composition	Soil	gram per kilogram (P)	g kg^{-1}
		percent (A)	%
Specific heat	Heat storage	joule per kilogram Kelvin	J kg^{-1} K^{-1}
Thermal conductivity	Heat flow	watt per meter Kelvin	W m^{-1} K^{-1}
Transpiration rate	H_2O flux density	gram per square meter second (P)	g m^{-2} s^{-1}
		cubic meter per square meter second (A)	m^3 m^{-2} s^{-1} or m s^{-1}
Volume	Field or Laboratory	cubic meter (P)	m^3
		liter (A)	L

(continued on next page)

Table 6. Continued.

Water content	Plant	gram water per kilogram wet or dry tissue	g kg^{-1}
	Soil	kilogram water per kilogram dry soil (P)	kg kg^{-1}
		cubic meter water per cubic meter soil (A)	m^3 m^{-3}
X-ray diffraction patterns	Soil	radians (P)	θ
		degrees (A)	°
Yield	Grain or forage yield	gram per square meter (P)	g m^{-2}
		kilogram per hectare (A)	kg ha^{-1}
		megagram per hectare (A)	Mg ha^{-1}
		tonne per hectare (A)	t ha^{-1}
	Mass of plant or plant part	gram (gram per plant or plant part)	g (g plant^{-1} or g kernel^{-1})

Table 7. Factors for converting non-SI units to acceptable units.

Non-SI Units		Acceptable Units
Multiply	By	To obtain
acre	4.05 × 10^3	square meter, m^2
acre	0.405	hectare, ha (10^4 m^2)
acre	4.05 × 10^{-3}	square kilometer, km^2 (10^6 m^2)
Angstrom unit	0.1	nanometer, nm (10^{-9} m)
atmosphere	0.101	megapascal, MPa (10^6 Pa)
bar	0.1	megapascal, MPa (10^6 Pa)
British thermal unit	1.05 × 10^3	joule, J
calorie	4.19	joule, J
calorie per square centimeter minute (irradiance)	698	watt per square meter, W m^{-2}
calorie per square centimeter (langley)	4.19 × 10^4	joules per square meter, J m^{-2}
cubic feet	0.028	cubic meter, m^3
cubic feet	28.3	liter, L (10^{-3} m^3)
cubic inch	1.64 × 10^{-5}	cubic meter, m^3
curie	3.7 × 10^{10}	becquerel, Bq
degrees (angle)	1.75 × 10^{-2}	radian, rad
dyne	10^{-5}	newton, N
erg	10^{-7}	joule, J
foot	0.305	meter, m
foot-pound	1.36	joule, J
gallon	3.78	liter, L (10^{-3} m^3)

(continued on next page)

Table 7. Continued.

gallon per acre	9.35	liter per hectare, L ha^{-1}
gauss	10^{-4}	tesla, T
gram per cubic centimeter	1.00	megagram per cubic meter, Mg m^{-3}
gram per square decimeter hour (transpiration)	27.8	milligram per square meter second, mg m^{-2} s^{-1} (10^{-3} g m^{-2} s^{-1})
inch	25.4	millimeter, mm (10^{-3} m)
micromole (H$_2$O) per square centimeter second (transpiration)	180	milligram (H$_2$O) per square meter second, mg m^{-2} s^{-1} (10^{-3} g m^{-2} s^{-1})
micron	1.00	micrometer, μm (10^{-6} m)
mile	1.61	kilometer, km (10^3 m)
mile per hour	0.477	meter per second, m s^{-1}
milligram per square decimeter hour (apparent photosynthesis)	0.0278	milligram per square meter second, mg m^{-2} s^{-1} (10^{-3} g m^{-2} s^{-1})
milligram per square centimeter second (transpiration)	10 000	milligram per square meter second, mg m^{-2} s^{-1} (10^{-3} g m^{-2} s^{-1})
millimho per centimeter	0.1	siemen per meter, S m^{-1}
ounce	28.4	gram, g (10^{-3} kg)
ounce (fluid)	2.96 × 10^{-2}	liter, L (10^{-3} m^3)
pint (liquid)	0.473	liter, L (10^{-3} m^3)
pound	454	gram, g (10^{-3} kg)
pound per acre	1.12	kilogram per hectare, kg ha^{-1}
pound per acre	1.12 × 10^{-3}	megagram per hectare, Mg ha^{-1}
pound per bushel	12.87	kilogram per cubic meter, kg m^{-3}
pound per cubic foot	16.02	kilogram per cubic meter, kg m^{-3}
pound per cubic inch	2.77 × 10^4	kilogram per cubic meter, kg m^{-3}
pound per square foot	47.9	pascal, Pa
pound per square inch	6.90 × 10^3	pascal, Pa
quart (liquid)	0.946	liter, L (10^{-3} m^3)
quintal (metric)	10^2	kilogram, kg
rad	1.00	0.01 Gy
roentgen	1.00	2.58 × 10^{-4} C (coulomb) kg^{-1}
square centimeter per gram	0.1	square meter per kilogram, m^2 kg^{-1}
square feet	9.29 × 10^{-2}	square meter, m^2
square inch	645	square millimeter, mm^2 (10^{-6} m^2)
square mile	2.59	square kilometer, km^2
square millimeter per gram	10^{-3}	square meter per kilogram, m^2 kg^{-1}
temperature (°F − 32)	0.556	temperature, °C
temperature (°C + 273)	1	temperature, K
tonne (metric)	10^3	kilogram, kg
ton (2000 lb)	907	kilogram, kg
ton (2000 lb) per acre	2.24	megagram per hectare, Mg ha^{-1}

ABBREVIATIONS AND SYMBOLS

General

Using commonly accepted abbreviations and symbols saves space in the journals and saves time for the writer and reader. Excessive use of abbreviations, however, may confuse a reader with jargon and may give a choppy appearance to the printed page.

Rules for abbreviating and many accepted abbreviations are given in the *CBE Style Manual*. Note that very few periods are used. Rules in other manuals may be helpful, but the periods used with many of the abbreviations given there are not acceptable in society publications. It is best to avoid using abbreviations in titles and abstracts. In the abstract, any abbreviations that seem necessary to use should be defined just as would be done in the main text.

Common Abbreviations

Accepted abbreviations and symbols are listed in the *CBE Style Manual*. Additional useful points are as follows:

1. The SI units used with numerals are abbreviated.
2. The names of states, territories, and U.S. possessions should always be spelled in full when standing alone. When they follow the name of a city, use the two-letter U.S. Postal Service code form (Illinois, IL) with or without the zip code number.
3. The symbol % is used with arabic numbers. The symbol is not repeated with each number in a range or series. Do not use the word "percent" with a number.
4. Names of months accompanied by day and year are abbreviated, except May, June, and July. In text, the month should be spelled out when used alone, with only day or year, and at the beginning of sentences. The month should always be abbreviated in footnotes, tables, and references. See "Time and Dates," this chapter.
5. The initials and last names of authors are used in the reference list and in footnotes.
6. The abbreviation or symbol of a unit of measurement should be used only if a number precedes the unit. The same abbreviation or symbol should be used for singular or plural forms of the unit. At the start of a sentence, the unit of measurement that follows a spelled out number should also be spelled out. For example, "Fifteen liters are . . ." but ". . . 15 L is."
7. Recognized symbols for chemical elements should be used without identification.

CONVENTIONS AND STYLE

8. In a series of measurements, the unit should be given at the end, i.e., 2 to 10°C; 3, 6, and 8 m.
9. United States of America is abbreviated USA. United States also may be abbreviated USA. The abbreviation U.S. should be used only as a modifier, e.g., U.S. government. Well-known government units may be abbreviated, e.g., USDA-ARS, TVA, without spelling them out.
10. The Latin name of an organism should be spelled out the first time it is used in the abstract and in the paper. Thereafter, use only the first letter of the genus name and spell out the species name. If the genus and species name start a sentence, the genus name should be spelled out. These Latin names are underlined (italicized).
11. The abbreviations "Lat" and "Long" should be used in expressions such as 30° N Lat or 30° W Long. They are not needed when they are used together: 30°N°20°W.
12. In a series of symbols or measurements where the first item starts a sentence, spell out only the first item, e.g., Nitrogen, P, K . . .

The following list of additional abbreviations is not exhaustive but gives the abbreviations commonly used in society publications. Please note abbreviations are not allowed for publisher's names.

a.i.	active ingredient	diam	diameter
Abstr.	Abstract	dry wt	dry weight
Agric.	Agricultural or Agriculture	EC	Enzyme Commission
		ed.	Editor(s) or Edition
Agron.	Agronomy	Eq.	equation
Am.	America or American	Exp.	Experiment
ARS	Agricultural Research Service	Fig.	figure (number)
		g (italic)	gravity, centrifugal
ASA	American Society of Agronomy	h	hour
		Handb.	Handbook
avg	average	i.d.	inside diameter
CI	Cereal Investigation (no.)	Illus.	Illustrations
		Inst.	Institute
conc	concentration	Int.	International
Conf.	Conference	J.	Journal
Congr.	Congress	min	minute
Conserv.	Conservation	Monogr.	Monograph
Counc.	Council	Natl.	National
CSRS	Cooperative State Research Service	no.	number
		o.d.	outside diameter
CSSA	Crop Science Society of America	PI	plant introduction (number)
cv.	cultivar	Pub.	Publication
Dep.	Department	Publ.	Publisher(s)

Rep.	Report	USA	United States of America
Res.	Research		
s	second	USDA	United States Department of Agriculture
Sci.	Science		
Serv.	Service		
Soc.	Society	USEPA	United States Environmental Protection Agency
Spec.	Special		
SSSA	Soil Science Society of America	USSR	Union of Soviet Socialist Republics
Stn.	Station		
TVA	Tennessee Valley Authority	Vol.	Volume
		vs.	versus
Univ.	University	yr	year

SPELLING AND CAPITALIZATION

Webster's Third New International Dictionary of the English Language, Unabridged (Gove, 1964) or a later edition, if available, is the societies' primary guide to spelling, capitalization, and compounding. Use of another dictionary, such as *The Random House College Dictionary* (Urdang, 1972) is usually acceptable. The *CBE Style Manual* is a helpful reference, especially in specialized word spelling, and *The Chicago Manual of Style* covers the subject across a range of disciplines. Italics (underline in manuscript) are used for isolated foreign language wording likely to be unfamiliar to readers. Familiar words and scholarly abbreviations such as en masse, in vitro, in vivo, in situ, et al., e.g., i.e., or ca. are set in roman type and should not be underlined.

The following are common rules for capitalization (proper names, first word in sentence, etc.).

The first letter is capitalized for:

1. Regions, sections, or groups of sites commonly associated together, e.g., Corn Belt, North Central states, the South, the West, Midwest; but northern Iowa.
2. First letter of genus, family, and order but not species.
3. Trademarked names but not adjectives derived from them.
4. First word after colon if it begins a clause not logically dependent on the preceding clause.
5. Any title immediately preceding a name: President Doe. There are some exceptions, e.g., Vice-presidential candidate Smith.

Words derived from proper names but now in common usage are not capitalized, e.g., paris green, bunsen burner, petri dish. Do not capitalize names of grasses, e.g., bermudagrass, sudangrass; or seasons of the year, e.g., spring, summer, fall, winter. Do not capitalize titles when not preceding a name, e.g., asssistant professor, research agronomist, and editor-in-chief.

NUMERALS

Reported data should include no more significant digits than the precision of the experimental methods warrant. Often, more than three significant digits of data from agronomic research cannot be justified. An acceptable rule is to round treatment means to 1/10 of their estimated standard error. For example, if the estimated standard error is 1.43, the means should be rounded to the nearest 0.1, and if the standard error is 18.4, the means should be rounded to the nearest 1.0.

Arabic numerals are generally preferable to roman numerals. Commas are not used to separate numbers greater than three digits. In text, four digit numbers are written together: 1000. A space separates numbers greater than four digits either left or right of the decimal point in tables and text: 10 000 and 0.052 067. In tables, when four-digit numbers and numbers of more than four digits occur together in a column, a space is used to separate the four digit number: For example:

      10
    100
  1 000
 10 000
1 000 000

Dates, page numbers, percentages, time, and numbers followed by units of measure are expressed with numerals: 2%, no. 1, Exp. 3, 1 g, 5 s, Eq. [1]. A numeral is used for a single number of 10 or more, except when the number is the first word of the sentence, or when preceded by a capitalized noun, e.g., Table 1 and Chapter 1. Words should be used for numbers below 10. Numerals are used for numbers nine and below when two or more numbers are used with them and one is above nine: ". . 2, 5, and 20 pots were planted . . ." but ". . . a group of 12 plants was incubated at three temperatures." Ordinal numbers are treated like cardinal numbers: third, fourth, 33rd, 100th, except in references (e.g., 5th ed., 7th Congr.). Large numbers ending in zeros use a word for part of the number: 1.6 million (not 1 600 000) or 23 μg (not 0.000 023 g). A zero is used before decimal numbers less than 1.0: 0.1 and 0.5. Use the connecting word "to" rather than a dash in a range of numbers, except when the numbers are used in parentheses or in tables.

PUNCTUATION

Punctuation marks help to show the meanings of words by grouping them into sentences, clauses, and phrases. These marks must be used precisely if the reader is to understand exactly the intended meaning.

The ordinary rules of punctuation are adequate for all ASA, CSSA, and SSSA publications. Some of these are given in the *CBE Style Manual* and *The Chicago Manual of Style*.

A few rules that are frequently violated are the following:

1. Use a comma before "and" or "or" in a series of three or more items, e.g., 2, 7, and 10 s.
2. Do not use a comma in a date that gives only the month and year, e.g., May 1965, or when one day of the month is given, e.g., 14 May 1965.
3. Do not use any punctuation after short items in a vertical list. For example: 1. Stems
    2. Roots
    3. Leaves
4. Use single quotes around cultivar names the first time the names are introduced in the abstract and paper. Place punctuation outside of the single quote marks.
5. Place double quotation marks outside of commas and periods but inside semicolons, colons, and question marks.
6. When it is necessary to enclose material within other statements already in parentheses, generally use brackets. Example: ". . . as stated (Jones, 1983 [as quoted by Smith, 1984])." Two exceptions to the use of brackets within parentheses are allowed for society publications. Use brackets if necessary to enclose scientific names that already contain parentheses, e.g., ". . . soybean [*Glycine max* (L.) Merr.] has the . . ." The use of brackets can be usually avoided by using commas. Example: ". . . soybean, *Glycine max* (L.) Merr., has the . . ." The other exception is mathematical usage, e.g., $A = [(b + c) - d] + e$. Also, numbered equations are referred to in text as Eq. [1].
7. Use the apostrophe when adding s for plural only to avoid confusion, e.g., 1920s, ABCs, but M.A.'s.

COMPOUND WORDS AND DERIVATIVES (USE OF HYPHEN)

A word containing a prefix, suffix, or combining form is a derivative and is almost always written as one word. Compound words used to express an idea different from that expressed by the separate parts are usually written as one word. Hyphens are used to avoid a confusing sequence of letters, a confusing sequence of adjectives, a jumble of ideas, or possible confusion with a word of the same spelling without the hyphen. Comprehensive rules for compounding are found in dictionaries and other books of usage.

Most compounds and derivatives fall under these general rules:

1. Derivatives are usually written solid, e.g., nonadditives, nonsignificant, postdoctoral, preemergent, antiquality, reuse, clockwise. Use hyphens with prefixes to words that begin with a capital and sometimes in a few awkward combinations that

bring two vowels together, e.g., un-American, semi-independent. For correct spelling check *Webster's Third New International Dictionary of the English Language, Unabridged.*
2. Hyphenate a compound adjective when used before, but not after, the word it modifies, e.g., winter-hardy plant; it is winter hardy; well-known method; it is well known.
3. Use a hyphen after a prefix to a unit modifier, e.g., semi-winter-hardy plant; non-winter-hardy plant.
4. Use a hyphen in a compound adjective that includes a number, including when the adjective is an abbreviated unit of measure, e.g., 10-yr-old field, 6-kg yield, 4-mm depth.
5. Noun compounds are usually formed when the term is a unit of measure or has special meaning or when one of the words has lost its accent, e.g., light-year, northeast, pineapple.
6. Do not use a hyphen after an adverb ending in -ly as the first part of a two-word modifier, e.g., widely known fact.
7. Use a hyphen for noun-adjective expressions for clarity; e.g., "On a per-gram basis."
8. Use hyphens to join numbers and abbreviations to chemical names, e.g., trans-2-bromocyclopentanol.
9. Use a dash (–), rather than a hyphen, between components of a mixed chemical reagent when typing the manuscript, e.g., $HCl-H_2SO_4$.
10. Always include the ionic charge when ions are included in combined expression, e.g., NO_3^--N, NO_2^--N, NH_4^+-N.

TIME AND DATES

Society publications use the 24-h time system, which is indicated by four digits—the first two for hours and the last two for minutes. In this system, the day begins at midnight, 0000 h, and the last minute is 2359 h. Thus, 2400 h of 31 Dec. 1983 is the same as 0000 h of 1 Jan. 1984.

The following style is used with the day of the month first, then the month, followed by the year: 18 Dec. 1984; 4 July 1984; but 18 December.

Julian day is frequently misused in biological literature to describe the calendar day of the year. The term is really the number of days from 1200 h on 1 Jan. 4713 B.C. A better term to describe the calendar day of the year is "day of the year."

MISCELLANEOUS STYLE POINTS

The usage advocated in this list is not exhaustive but merely covers some recent problem areas.

Abstract—Footnotes or literature references may not be used in the abstract.

Affect, Effect—These words are often confused by authors. Affect is usually correct as a verb. Never use it as a noun or as an adjective. Effect is always correct as a noun but can also be used as a verb. Correct: "Light affects growth; light effects the conversion of chlorophyll to metastable products."

Commas—Use commas before changes of subject. Authors often use them when they are not needed. Try to break up nouns and clusters with commas when possible, e.g., "Dissolved in H_2O, NaCl forms a saline solution."

Cultivar Names—Use these names instead of repeating the genus and species or a long common name. Single quotes enclose the name the first time used. It is redundant to use the word cultivar and single quotes at the same time. Refer to *Registered Field Crop Varieties: 1926-1981* (CSSA, 1982) and other lists of published registrations for names of cultivars.

Foreign Numbers and Spellings—Use American spelling except for titles in references. Use sulfur not sulphur, unless in a citation.

Hardiness—Use Webster's: winterhardiness, winter hardy, winter-hardy plant.

However—Be careful of this word. It is most often used between commas as a transition word, but it may also be used as an adjective, e.g., ". . . however great the cost . . ." Use it rarely at beginning of a sentence.

Nitrogen Fixation—Biological fixation of atmospheric nitrogen is correctly termed "dinitrogen fixation" and is represented in abbreviated form as "N_2 fixation" or "N_2-fixing ability."

Number—This word can be abbreviated to no. in tables, but it is best spelled out in text. There are exceptions: In methods and footnotes, it is permissible to use plant no. 3 or no. 3 needle. It is proper to use arabic numbers in this case and when scoring systems are used: "a scale of 0 to 5" rather than "a scale of zero to five."

Number of Sets of Chromosomes—Use 2x, 3x. The x's are not times signs.

Ranges—Use the connecting word "to" rather than a hyphen: -22.9 to $14.9°C$. Don't repeat °C. If the range is given in parentheses or in a table, use a dash.

Rhizobium, rhizobia, rhizobial—The word *Rhizobium* (italics) is a genus name that has only one meaning: the proper noun referring to the genus of bacteria. The collective noun "rhizobia" means more than one cell of *Rhizobium*. The adjective "rhizobial" describes some attributes of *Rhizobium*.

Subscripts, Superscripts, Overscores, Accented letters—Use these marks with discretion because they are difficult to typeset when several characters are used both as superscripts and subscripts.

Was, Were—Use "a total was," "none was," "data were," and "either was."

Which, That—"Which" introduces a nonrestrictive clause and is usually preceded by a comma (I do not drink sea water, which is salty.). "That" introduces a restrictive clause (I do not drink water that is salty.).

CHAPTER 7

Tables, Illustrations, and Mathematics

Tables and illustrations are used to support conclusions or illustrate concepts. Each figure or table should be entirely informative in itself. Text should be written around the tables and figures rather than appended to them. Tables and figures should be planned early in the writing of the manuscript and a minimum number of words used to describe them.

Tables and illustrations have essential differences in purpose. Tables present accurate numbers for comparison with other numbers. Illustrations reveal trends or record natural appearance. Sometimes these purposes overlap, but they rarely substitute for one another. Data presented in tables should not be duplicated in figures.

The rest of this chapter emphasizes technical preparation of tables, illustrations, and mathematical equations for society publications.

When needed, using multiplication factors after units in tables or figures means that the measured data have been multiplied by that factor to result in the presented data. Avoid use of multiplication factors by selecting the proper prefix as discussed in the SI units section.

TABLES

Tables are used for reporting extensive numerical data in an organized manner. They should be self-explanatory. It is seldom necessary to use a table for fewer than eight items of data.

Table captions should be brief but sufficiently explain the data included. They *should not* include the units of measurement. Number the tables consecutively and refer to them in numerical order in the text as Table 1, Table 2, etc.

The principal parts of a table are shown on p. 62. Follow this general form for the stub and field items. Show the units for all measurements, in spanner heads, in column heads, or in the field. In general, only horizontal rules are used: a single rule at the top, a single rule

		Spanner head (if applicable to all columns)‡		
		Subspanner head§		
Stub head†	Column head	Column head¶	Column head#	Column head††
unit	unit	———— unit ————		
(Stub)		(Field)		
		Independent line‡‡		
Main entry line				
Subentry line§§			536**	
Subsubentry line		208¶¶		
Subentry line				105*
		Independent line##		
Main entry line†††				
Subentry line				

*,** Significant at the 0.05 and 0.01 probability levels, respectively.
†, ‡, §, ¶, #, ††, ‡‡, §§, ¶¶, ##, ††† Supplementary notes.

below the boxhead, and a single rule at the bottom just over the footnotes. Additional horizontal rules may be needed under spanner heads and subheads.

Two types of footnotes are used with tables: those to show statistical significance and those to give supplementary information. The * and ** are always used in this order to show statistical significance (or nonsignificance) at the 0.05 and 0.01 probability levels, respectively, and cannot be used for other footnotes. Significance at other levels should be designated by a supplemental note. Lack of significance at any level is usually indicated by NS. Supplementary notes are given the following symbols in this order: †, ‡, §, ¶, #, ††, ‡‡, etc. These symbols should be cited just as you would read a table—from left to right and from top to bottom. When asterisks are part of a table, they precede the other notes.

Draw symbols by hand, if necessary. Do not use numbers or letters.

Numbers with the same unit and/or equal length should be centered in the column. If they are unequal, center the longest one and align the rest on the decimal point. If a column contains various units, like units should be aligned, but different units may be centered differently.

Indicate italics by underlining if italic lettering is not available on the typewriter used. The manuscript for tables should be double spaced. Type each table and its heading on a separate sheet. Additional information on tables is given in the *CBE Style Manual*.

FIGURES (ILLUSTRATIONS)

Figures are often the best means for presenting scientific data, but they are expensive to prepare and publish; therefore, each one should tell its "thousand words" or be omitted. Figures are made from drawings or from photographs or other shaded materials. The preparation of these two types of figures is discussed below. The managing editor will reject a drawing or photo not suitable for acceptable reproduction in the publication.

A figure caption should be brief but sufficiently explanatory of the data included to tell its own story. Identification of the curves or other parts should be included in the figure itself when possible. Refer to figures as Fig. 1, Fig. 2, etc. in the captions and in the text. Type all captions together on a separate sheet. Draw or mount each figure on 216 by 280 mm (8½ by 11 inch) paper. Number each figure in the margin, and include author's name on each one. Additional suggestions on preparing figures may be found in the *CBE Style Manual*.

PHOTOGRAPHS (SHADED MATERIAL)

Authors should keep in mind that a photograph should be used only if it shows something essential for the point being made. Photos made for slide projections or talks are rarely usable for articles. Photographs should be examined to determine whether each shows something unique, interesting, and clearly identifiable.

Only high-quality, black and white glossy prints with sharp details and good contrast should be used. Prints should be in sharp focus, and have good density. The following suggestions, if followed, should help authors achieve better photographs and photomicrographs in published articles:

1. In taking photographs, provide a point of reference for the reader. A field of corn may look like any other, but a specific focal point can illustrate the significance of the message you are trying to convey. For example, a ruler is a simple point of reference in a photograph describing the effect of several chemical herbicides on plant height. Indicate the magnification of photomicrographs in the legend, include a bar indicating scale, or do both. Eliminate extraneous matter by special care in taking the picture, by careful cropping of the negative or print, or by use of a spray gun on unwanted areas of the print.
2. To provide clear contrast in the print, take photographs under the proper light conditions and with the proper background.
3. When photos are taken in a series, use the same camera angle and distance from the subject. A picture taken at 3 m from the subject at 0800 h will be quite different from a picture taken at 6 m at 1700 h.
4. When making prints or photographs, keep in mind the final column width of the publication. Scientific journals use a single column width of 84 mm (3¼ inches) and a double column

width of 178 mm (7 inches). *Crops and Soils Magazine* uses a one, two, and three column width of 53 mm (2 inches), 112 mm (4½ inches), and 178 mm (7 inches), respectively. Books have a column width of 110 mm (4½ inches). Occasionally photographs can be placed lengthwise on the page, but the practice is avoided, if possible. Most photographs benefit from some reduction in size in the final printing, except for electron micrographs, which reproduce best if they are not reduced.

5. Provide glossy prints of approximately 200 by 250 mm (8 by 10 inches); if possible, small photomicrographs should be mounted.
6. If letters or numbers on the photographs are used, make them large enough to withstand reduction. Be sure that letters and numbers in a series of prints are all of uniform height and density, clear, and legible. Equally important, for photomicrographs, is the necessity of having contrast for any written material added to the subject of the picture. If the background is black or nearly so, use white lettering. In many instances, the numbers, letters, scales, and arrows that depict something specific in the photograph are lost against the background. One solution to this problem is to place a solid white circle or square behind the letter, number, or scale. See next section on graphs and charts.
7. Carefully crop a photograph or indicate where you wish it cropped. A photograph with 75 mm (3 inches) of horizon uses much needed space.
8. Except for series shots, take several pictures, from different angles and under different lighting conditions, but always use the best ones. Keep the rest until publication because the editor may ask for them.
9. Carefully save the negatives from one of a kind photographs. The societies do not ordinarily return prints.
10. Obtain good quality prints from the photo shop. Do not accept underexposed and overexposed or underdeveloped and overdeveloped prints. They give poor-quality halftone reproduction in the printing process. Touching up prints usually indicates below average quality. Obtain a second print.
11. When two or more photographs are to be combined into one photograph, mount the parts together on paper or lightweight cardboard. Be sure that each part of the copy is trimmed square and that all the parts lie flat and fit snugly together without excess paste or other mounting material around the margins. Be sure that each part of a composite figure is clearly identified on the figure or in the caption by letters, e.g., A, B, . . .
12. If you must write on the back of the photograph, do it with a soft lead pencil. Supply the name of the author, short name of the article, and number for each figure. Send photographs

with cardboard around them to minimize possible damage due to mailing and handling. If possible, mount each photograph separately on cardboard, with mounting corners, so they can be removed easily.
13. If a person or named product is shown in a photo, the author must obtain written permission from the person or the manufacturer of the product for use of the photo. The society is not responsible for any claims that may result.
14. Photographs of other photographs are unacceptable.

GRAPHS AND CHARTS (LINE DRAWINGS)

Graphs and charts should be designed to improve the general presentation of a technical publication by reporting data in an easily comprehensible manner. Those preparing drawings for society publications should read the suggestions presented here and study the sample graphs at the end of this section.

The decision to select and use charts or graphs should be governed by the writer's message and the points to be brought out in the illustration. Graphs primarily show trends. It is not necessary to show all the coordinate ruling in most graphs. Satisfactory results will be obtained by using tracing cloth or plain white paper, with only the necessary vertical and horizontal lines drawn in.

Just as in the preparation of photographs, the size of the final graph or chart is important.

All figures in ASA, CSSA, and SSSA journal articles are made to conform to a column width of approximately 85 mm (3.5 inches) or a full-page width of approximately 178 mm (7 inches) of the journal. Most photographs and drawings are reduced 40 to 60% in making the negatives for printing. The size of the original figure and the lettering on it should be appropriate for the required reduction—to the width of one column or less in most cases. The page widths of other publications are given in the previous section on photographs. Keep these dimensions in mind when preparing graphs and charts or give the dimensions to the artist who is executing the material.

If various lines in the graph are identified by different markings or symbols, these should be identified within the graph rather than in the captions. The symbols should be a large enough size to withstand reduction of the figure to the final size of publication and still be legible. Abbreviations should follow the same style used in the text.

To promote uniformity in lettering in the journal illustrations, authors and artists should use templates and lettering sets to make the lettering uniform on their graphs and charts. The proper sizes of letters are discussed below.

Glossy photographs or photocopies of charts or graphs may be submitted for review in place of the original drawings. Once the paper is approved for publication, the original drawings or good glossy photographs must be sent to headquarters office. Photocopies will be unacceptable. These must be sharp and clear in all details and should be at least 130 by 178 mm (5 by 7 inches).

Photos smaller than 130 by 200 mm (5 by 8 inches) should be mounted on manuscript-size paper. Do not have drawings reduced to the size desired in the printed form; better printing results when the drawings are reduced at headquarters.

PREPARING THE DRAWING

A high degree of uniformity is desired in drawings for the journals in terms of style of the letters and finished size of the letters. Uniformity is especially important in all drawings prepared for a single manuscript.

Size of Original Drawing

Plan for a net reduction of 40 to 60% of the original size. For a one-column figure this means that the original drawing should be 178 to 210 mm (7 to 8½ inches) wide; for two-column figures the original drawing can range from 360 to 440 mm (14½ to 17½ inches) wide.

Drafting Materials

Use white drawing paper or tracing vellum and black india ink.

Thickness of Ruled Lines

Use a medium width line, e.g., the no. 2 point, in pens such as Leroy, Rapidograph, and Castell, for the graph borders or reference lines. For lighter lines (grids, bonds, arrows) use the no. 1 point and for heavier lines (graph curves), use a no. 2½ or no. 3 point. Do not use extremely fine lines (no. 0 or 00 points) for any part of the drawing or lettering.

Height and Style of Letters and Numbers

Height

The desired minimum height for numerals and capital letters in the finished journal or book page is 1.5 to 1.75 mm (7- to 8-point type size); for lowercase letters and symbols, it is 1.25 mm (6 point type).

If the drawing is to reproduce at 40 to 50% of the original size, the height of the numerals and capital letters on the drawing itself should be 4 to 5 mm; the height for lowercase letters should be 3 to 3.5 mm.

When using dry-transfer or pressure-sensitive letters, use sheets identified as 18-point or 24-point for the main numerals and headings, and 14-point for the subsidiary items, if any.

Style of Lettering

Use open style or block letters (usually referred to as sans serif). In the pressure-sensitive letters, the Univers no. 55 typeface in 18- or 24-point (or in some cases the 14-point) is an excellent all-purpose style of lettering for graphs and drawings. Helios is also recommended. CHARTPAK is one of the trade names of this art material. See samples below.

For mechanical lettering sets such as Leroy, Rapidograph, and Wrico, the Leroy templates no.'s 140, 175, and 200; the Rapidograph scriber guides CL140, 175, and 200, or the Rapidoguide no. 3031 or 3030 will give the desired style and height of lettering (between 3.5 and 5 mm).

When using these mechanical lettering guides use the no. 1 pen size.

Capital or Lowercase Letters

Use of all capital letters is preferred for the main axis labels, main graph lines labels, and other main items in the illustration. Use lowercase letters, however, for units of measure that follow the main label.

For secondary headings within the illustration, use capital and lowercase letters (each main word starts with a capital letter) of the same letter size as in the all capital labels. That is, if 18-point letters are used for the all capital letters, use 18-point letters for the capital and lowercase letters.

If additional subsidiary labels are needed in the illustration, use all capital letters in the next lower letter size. That is, if 18-point letters are used for the first- and second-level labels, use 14-point all capital letters for the third-level labels. Do not make them less than 12-point letters or 2.5 mm.

Mechanical Lettering Guides

RECOMMENDED TEMPLATE & PEN COMBINATIONS IN ACTUAL SIZE

The chart shows the wide range of LEROY lettering effects that can be produced by combining various LEROY templates and pens.

*These template and pen sizes should be your first choice.

Capital letter height, mm	Template size	Pen size 0	Pen size 1	Pen size 2	Pen size 3
2.0	80	C			
2.5	100	B	C	D	
3.1	120	C	D	E	
3.6	140*	C	D*	E	F
4.5	175*	C	D*	E	F
5.1	200*	C	D*	E	F
6.1	240	C	D	E	F

Dry Transfer of Pressure-Sensitive Letters

UNIVERS 55 *All lettering shown actual size*

14 POINT —
ABCDEFGHIJKLMNOPQRSTUVWX
abcdefghijklmnopqrstuvwx   1234567890

18 POINT —
ABCDEFGHIJKLMNOPQR
abcdefghijklmnopqr   1234567890

24 POINT —
ABCDEFGHIJKLM
abcdefghijklm   12345678

Examples of Acceptable and Unacceptable Figures

The examples the following pages, composites of actual material received, show acceptable and unacceptable figures.

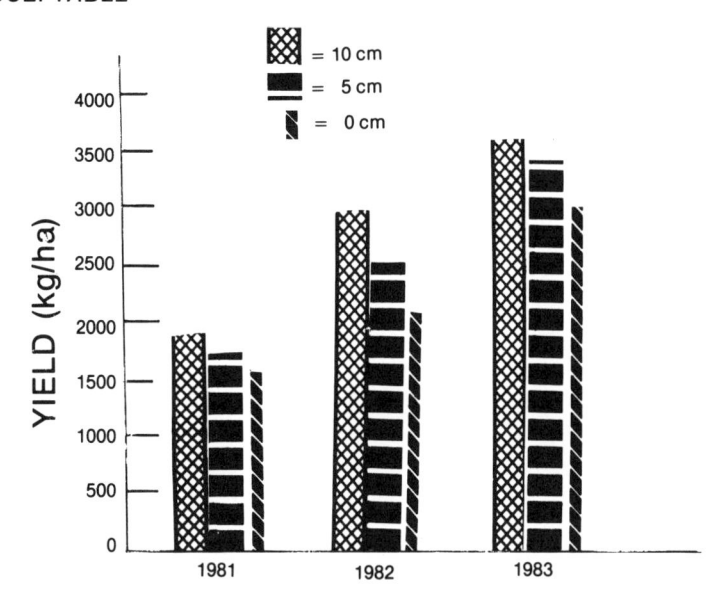

UNACCEPTABLE

[Bar chart showing YIELD (kg/ha) vs DEPTH OF P PLACEMENT (cm) for years 1981, 1982, 1983 with patterns for 10 cm, 5 cm, and 0 cm]

Complicated patterns and small (dates, numbers, above) type are difficult to reproduce in journal size.

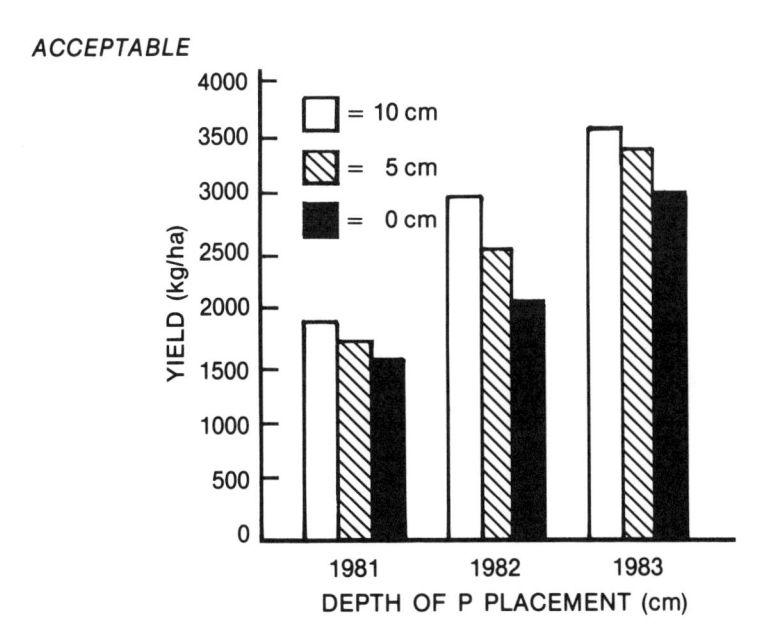

ACCEPTABLE

[Bar chart showing YIELD (kg/ha) vs DEPTH OF P PLACEMENT (cm) for years 1981, 1982, 1983 with simple patterns for 10 cm, 5 cm, and 0 cm]

Use simple patterns and the same type size for all parts of the figure. Make sure the type is the correct size.

UNACCEPTABLE

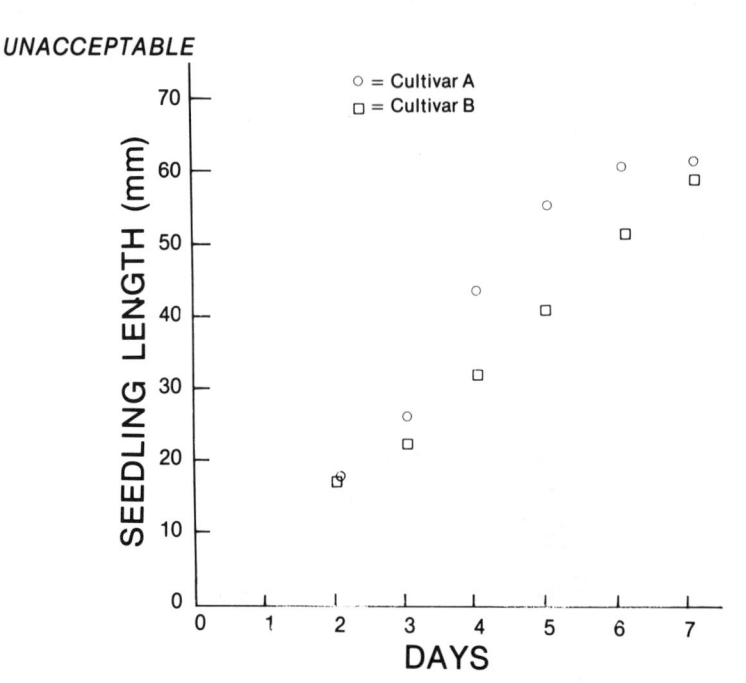

Small symbols, thin lines, and inefficient use of space complicate production as well as comprehension.

ACCEPTABLE

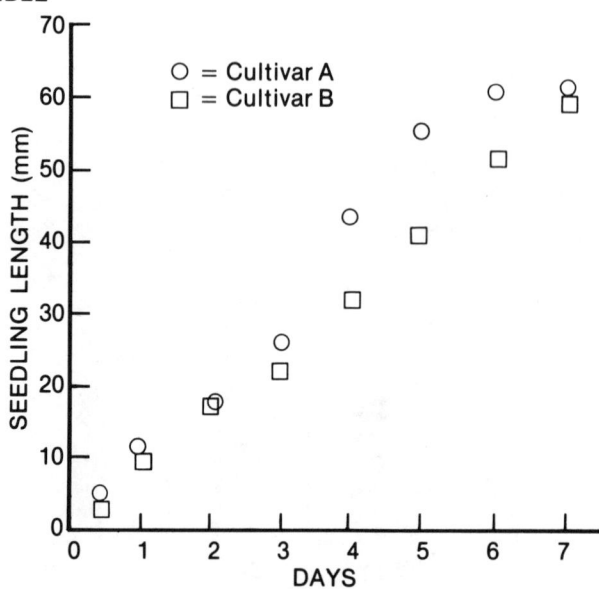

Make efficient use of space in graphs by starting and ending the axes at the appropriate place. Use medium width lines, choose symbols large enough to withstand reduction to publication size, and affix correct and consistent legend and label type size.

Additional Important Suggestions

1. Never use a typewriter or hand-drawn letters for illustrations.
2. Be sure that the decimal points are correctly positioned and large enough to stand reduction in size. The size of the decimal point should be in proportion to the accompanying numbers. Place the decimal points at the base of the numbers, not centered halfway up as is done in some European journals.
3. Use sharp glossy photographs for drawings if the originals are filed or if the original is very large. Be sure the photos are 200 by 250 mm (8 by 10 inches) and that no distortion or out of focus areas have been allowed in the photo printing process.
4. Clearly label all illustrations along the top or bottom margin indicating author and the figure number. Also, indicate which edge is the top of the illustration.
5. Computer-drawn graphs are acceptable for society publication if they meet the standards discussed above. Most problems in reproduction lie in uneven type and broken lines. Make all lines, letters, and numbers with continuous solid black lines. Consult your computer programmer for proper combination of type size and density of characters on the printout.
6. Use the same abbreviations for units of measurements in the figures as were used in the text.
7. Do not send photographs or drawings that require enlarging. If this is necessary, have the enlargement made locally and check it for quality before deciding on its use.
8. Make lines straight and corners square.
9. Use solid black, white, diagonal lines, dot screen, or a random dot for bar graph patterns. Do not use decorative borders.

MATHEMATICAL EQUATIONS

Mathematical equations give journals and books some of their most difficult and costly problems of type composition. If not clearly and unmistakably marked in the manuscript, they cause errors from the copy editor to the compositor, the proofreader, the author, and back to the editor. What may be perfectly clear to the author can be bewildering to anyone who is not a mathematician. Keep in mind that typesetters will reproduce what they see, and not what you know. Therefore, preparation of the manuscript copy and all directions and identification of letters and symbols must be clear, so that those lacking in mathematical background can follow the copy.

Position and Spacing

Positions, spacings, and other details must be exactly as they are to be in printed form. If your typewriter does not have special symbols, draw them neatly by hand. Draw or type superscript and subscript letters and symbols in the correct positions, and mark them in pencil if there could be any confusion about them, e.g., e^x $\log_{10}$ B_A. The typist must put the proper space between symbols. Use a hyphen with spaces on both sides for the minus sign, but omit the space after the hyphen when the sign represents a negative quantity.

Identifying Characters

Call attention in pencil marks to obscure modifications of symbols, e.g., prime marks, dots over symbols, and similar situations. Carefully distinguish between the letter "O" and zero; the letter "I", lowercase L, and the number "1"; and the degree symbol and the superior letter "o" or zero. When "x" represents the multiplication sign, indicate this lightly in pencil marks or write "multi. sign" above it. Indicate desired type size and face as follows: capitals—triple underline, small capitals—double underline, italic type—single underline, and boldface type—wavy underline. Identify these characters in text as well as in equations and tables.

Fractions and Simplifying Equations

Use slant rule fractions (e.g., x/g) as much as possible, especially in the text. Show the necessary aggregation by using parentheses, brackets, and braces. Use the sequence { [( )] }.

In horizontal rule fractions, be sure to align the rules with the main signs of the equation or formula. In complex equations, use horizontal rules for the main fractions and slant rules in numerators, denominators, and exponents.

Use	Instead of
a/bcd	$\dfrac{a}{bcd}$
$a/(b-c)$	$\dfrac{a}{b-c}$
$(a/c)-(b/d)$	$\dfrac{a}{c}-\dfrac{b}{d}$

Exponential Functions

For exponentials with lengthy or complicated exponents, the symbol exp should be used, particularly if such exponentials appear in the body of the text. Thus, exp $(a^2 + b^2)^{1/2}$ is preferable to $e^{(a^2+b^2)^{1/2}}$. The larger size of symbols permitted by this usage also makes reading easier. In index size the structure of some of the complicated exponents may entail expensive hand setting.

Integral Signs and Limits

With single integral signs, the upper and lower limits should always be placed to the right of the integral sign, never above and below.

$$\int_z^a \quad \text{not} \quad \int\limits_z^a$$

Radical Signs

Use fractional and negative exponents, wherever possible, instead of the radical sign and fractions. The following example illustrates conversion with the fractional exponent and also the saving in space which may be made with the use of the solidus or slant line.

$$\frac{\cos\frac{1}{x}}{\sqrt{a+\frac{b}{x}}} \quad \text{should be written} \quad \frac{\cos(1/x)}{[a+(b/x)]^{1/2}}$$

If the radical sign is necessary, use the partial rather than the complete sign, e.g., $\sqrt{\,}(x/g)$ instead of $\sqrt{(x/y)}$.

Accented Letters and Symbols

Some accented letters are a typesetting problem such as accents on Greek letters or lowercase letters with ascenders (e.g. $\hat{\sigma}$, $\hat{b}$). Avoid wide use of symbols or letters with accents or bars placed directly above or below. Try to use primes, suffixes, or asterisks, e.g., use a' or a^* instead of $\breve{a}$, $\bar{a}$, or $\ddot{a}$. To designate "average," use pointed brackets instead of an overbar; <a + b>, not $\overline{a+b}$. Spell out "mean" instead of using $\bar{x}$.

Numbering Equations

It is not necessary to number all displayed equations. Usually only those that are referred to elsewhere are numbered. If equations are numbered, place the numbers in brackets at the right margin. Refer to them in the text as Eq. [1], Eq. [10], etc.

Additional References for Mathematics

More information on rules and suggestions in preparing mathematical copy can be found in the *CBE Style Manual* and in other references listed in the bibliography.

Greek Letters

Unless special typewriter keys are available, a fine-pointed pen or pencil should be used for making Greek letters. These letters should be identified in the margins the first time they appear and should be marked whether capital or lowercase, e.g., "cap. beta" or "l.c. sigma."

The Greek alphabet, showing both capital and lowercase letters, is given below. Modifications of a few of these letters may be acceptable, but the ones given here should be used insofar as possible.

	Capital	Lower case		Capital	Lower case		Capital	Lower case
alpha	A	α	iota	I	ι	rho	P	ρ
beta	B	β	kappa	K	κ	sigma	Σ	σ, ς
gamma	Γ	γ	lambda	Λ	λ	tau	T	τ
delta	Δ	δ, ∂	mu	M	μ	upsilon	Υ	υ
epsilon	E	ϵ	nu	N	ν	phi	Φ	ϕ, φ
zeta	Z	ζ	xi	Ξ	ξ	chi	X	χ
eta	H	η	omicron	O	o	psi	Ψ	ψ
theta	Θ	θ, ϑ	pi	Π	π	omega	Ω	ω

CHAPTER 8

Proofreading

Galley, figure, and table proofs, along with manuscripts for all articles, are sent to the authors for review before the articles are published. The authors are responsible for the careful proofing and prompt return of the proofs and manuscripts. They should answer all questions found in memos from the editors and those marked in the margins of the proofs or manuscript.

The editors, typesetters, and printers exercise great care in avoiding and correcting errors, but they cannot assume responsibility for any that are not marked by the author on the proofs. The author is entirely responsible for the correct spelling of names and for other information given in citations and for citing the references in the proper places in the text. Authors are also responsible for the accuracy of all facts, dates, statistics, and the position of all parts of mathematical formula.

Keep in mind that the proof is not the manuscript. Corrections in the manuscript copy are made within the line at the point of correction. The compositor then reads along line by line and sets the type while reading. But once the type has been set and the copy is in galley proof form, the editor and compositor will no longer read each line to find where changes have been made. Instead, they will look down the margins of the galley proofs to find the appropriate proofreader's mark opposite the line where a correction is needed. Therefore, mark all corrections in the margin of the galley proof.

Read the galley proof twice—first with a person reading from the manuscript to avoid omissions of full lines or paragraphs, errors in dates, statistics, and mathematical formula. Check all references to tables, figures, and literature citations. Read the proof a second time alone. Be careful that your eyes do not pass over misspellings and omissions.

A sample corrected proof and some common proofreader's marks are shown on the next pages. Study the sample galley and the proofreader's marks carefully before you check your proofs. Keep the following points in mind (see also *CBE Style Manual*, p. 105-111):

   1. Use red ink or pencil on the proof to indicate printer's errors, and blue for your alterations. Take care to write legibly.

2. Place all correction marks in the margins of the proofs, either left or right, opposite the line in which the error occurs. If there are a large number of corrections in one line or in an adjoining line, use both left and right margins for marking.
3. Make two marks on the proof for each correction: one or more in the text line using a caret and short perpendicular line (λ) to show *where* he change is needed, and the other in the margin to show *what* change is needed.
4. Cross out the unwanted characters and put a delete sign in the margin when material is to be deleted and nothing added in its place. When material is to be substituted for a deletion, don't use the delete sign; just cross out the unwanted material and write the new copy in the margin.
5. If there are several corrections in one line, separate one from another by slant lines (e.g., λ /☉/☉) and arrange them in order to read from left to right. If the same correction is made in two or more places in the same line or in adjoining lines, write the correction once and follow it with a number of slant lines to equal the number of corrections.
6. Type insertions of more than one line at the bottom of the galley proof or type them on a separate sheet and attach to the galley with tape. Show clearly where new copy is to be inserted. All changes must be on the galley proof and not in a separate letter or note.
7. If a word, phrase, or sentence is substituted in a paragraph, make the substitution as nearly as possible the same number of letters as the deleted matter. If possible, delete at the end of a paragraph rather than in the middle or at the beginning.
8. Transposing words, phrases, or sentences is costly. Transposing paragraphs is relatively inexpensive.
9. Always make changes or answer questions on the galley proof, *never* on the *dead* manuscript.
10. Look for question marks on the manuscript and proofs. Give a clear answer to every query from the editor or typesetter.
11. Only essential changes should be requested. Authors of journal articles will be charged for revisions or additions to the proofs that are changes from the original manuscript. Author errors or additions should be indicated on the proof with a circled "AA."
12. Printer's errors (deviations from the original copy) are corrected at no cost to the author. Any printer's errors detected should be indicated on the proof with a circled "PE."
13. Indicate the position of figures and tables by a marginal notation with a circle around it; also encircle all notes, etc. not to be set in type.
14. Proofs of your tables and figures in the final size are included with the galley proofs. The table proofs should be carefully

proofread; mark corrections directly on the proofs. Check the figures to be sure they are accurate and correctly identified.
15. The desired return date for the proofs is indicated. Use airmail for locations outside the USA and first-class mail in the USA.

When proofs are returned, the production editor looks them over carefully, edits the changes or additions if necessary, and transfers the corrections to another set of proofs. The corrections are made to the typeset material prior to page make up.

SAMPLE GALLEY PROOF

Mean symptom readings (Table 1 showed that Avena sterilis lines were more resistant than the A. sativa checks and that the resistance was transmitted to F₁ and F₂ hybrids with Lamar. The dominance effects were consistently smaller than the cumlative additive effects and did not differ significantly from zero in six cases out of eight (Table 1). although these results indicate that inheritance of resistance was determined predominantly by additive gene action, they should not be taken as proof that dominance was not involved. In fact, the performmances of the F₁'s (Table 1, Fig 1) were not at the midpoint between those of their parnts but were closer to those of the resistant A. sterilis parental lines, suggesting some dominance for resistance.

PROOFREADER'S MARKS

∧ Caret—something to be inserted; mark in text line

⊙ Period

⋀ Comma

⊙ ⁄ ⁄: Colon

;/ Semicolon

⋎ Apostrophe

⋎/⋎ Quotations

=/ɑ/H Hyphen

Mark	Meaning	Mark	Meaning
(/)	Parentheses	*less* Ⓢ	Less space
⊏/⊐	Brackets	Ⓢ	Insert space
□	Indent one em, double for two em, and so on	*eq* Ⓢ	Equalize spacing
1/M	One em dash	ͻ	Turn letter or line
1/N	One en dash (short dash)	[or]	Move to left or to right
⌒	Close up	⊓ or ⊔	Move up or move down
stet	Let it stand, when something has been inadvertently crossed out. Dots under matter will usually suffice, but also include "stet" on margin to avoid misunderstanding.	*tr*	Transpose
		(b\a) *tr*	Character to go around letters, words, or phrases to indicate that they are to be transposed. Always include "tr" on margin of proof.
ͽ, ⌐	Delete—take out	*lc*	Lower case
⌐	Delete and close up	\2/	Superior letter or figure
×	Broken letters or defective type	/a\	Inferior letter or figure
		———	Italics and *ital*
¶	Paragraph	*rom*	Roman
no ¶	No paragraph	*bf*	Boldface
wf	Wrong font	⑦	Circle around figures means spell out
≡	Capitals and *caps*	(twenty grams)	Circle around word means use figure or abbreviation.
=	Small capitals and *sc*		

Correction requires two marks on the proof—one in the margin indicating what is to be done and another within the type indicating by a short perpendicular line or a caret (‸) the exact place where the change is to be made.

CHAPTER 9

Copyright and Permission to Print

To comply with the provisions of the U.S. Copyright Law of 1978 (P.L. 94-553), the societies handle copyright and permissions in the following ways.

1. A Permission to Print and Reprint form is used when the societies do not intend to copyright an individual article in a publication. Authors of such articles, when signing the form, still retain the authority and responsibility to decide upon and respond to requests for further use of the article by other persons or organizations.
2. A Transfer of Copyright form is used for publications where the individual articles are copyrighted by the societies. Details of the transfer agreement are given on the form.

Each of the above forms has two check items, one for U.S. government authors and the other for nongovernment authors. Generally, work done on government time is in the public domain and cannot be copyright, but the form certifies how the work was done.

The copyright law also requires that permission be obtained to use material published elsewhere. It is the author's responsibility to obtain permission from the owner of the material if it is not in the public domain. A letter should be sent requesting permission. The signed letter granting permission should be attached to the manuscript so that it can be permanently filed in the ASA Headquarters Office at Madison, WI.

Name and address of
copyright owner

Dear

    I am writing an article entitled _____
_____ .
to be published in _____ .
    I request your permission to include in my article the following material _____
_____

Volume _____ Page(s) _____ Year _____ from the article _____
_____
written by _____ .
    If you grant your permission, please sign in the space below and return this letter to me. Thank you.

                              Sincerely,

Permission to use the above-cited material in the publication described above, and the subsequent reprints, editions, and translations of it in a nonexclusive manner is granted, provided proper credit to the author and publisher is made.

By _____ _____
               Print Name   Sign Name

Date _____ Title _____

Sample letter for requesting permission to reproduce material from another source.

CHAPTER 10

Publication Title Abbreviations

Adv. Agron.	Advances in Agronomy
Agric. Eng.	Agricultural Engineering
Agronomy	Agronomy (Monographs)
Agron. J.	Agronomy Journal
Am. J. Bot.	American Journal of Botany
Am. Mineral.	American Mineralogist
Am. Soc. Test. Mater.	American Society for Testing and Materials
Anal. Chem.	Analytical Chemistry
Ann. Appl. Biol.	Annals of Applied Biology
Ann. Bot.	Annals of Botany
Annu. Rev. Microbiol.	Annual Review of Microbiology
Aust. J. Agric. Res.	Australian Journal of Agricultural Research
Aust. J. Plant Physiol.	Australian Journal of Plant Physiology
Biochem. J.	Biochemical Journal
Bot. Gaz. (Chicago)	Botanical Gazette
Can. J. Genet. Cytol.	Canadian Journal of Genetics and Cytology
Can. J. Plant Sci.	Canadian Journal of Plant Science
Can. J. Soil Sci.	Canadian Journal of Soil Science
Clay Miner. Bull.	Clay Minerals Bulletin
Crop. Sci.	Crop Science
Crops Soils	Crops and Soils Magazine
Discuss. Faraday Soc.	Discussions of the Faraday Society
Environ. Sci. Technol.	Environmental Science and Technology
Ind. Eng. Chem.	Industrial and Engineering Chemistry
J. Agric. Food Chem.	Journal of Agricultural and Food Chemistry
J. Agron. Educ.	Journal of Agronomic Education

J. Am. Soc. Sugar Beet Technol.	Journal of the American Society of Sugar Beet Technologists
J. Atmos. Sci.	Journal of Atmospheric Sciences
J. Appl. Bacteriol.	Journal of Applied Bacteriology
J. Appl. Phys.	Journal of Applied Physics
J. Assoc. Off. Anal. Chem.	Journal of the Association of Official Analytical Chemists
J. Biol. Chem.	Journal of Biological Chemistry
J. Br. Grassl. Soc.	Journal of the British Grassland Society
J. Colloid Sci.	Journal of Colloid Science
J. Econ. Entomol.	Journal of Economic Entomology
J. Environ. Qual.	Journal of Environmental Quality
J. Hered.	Journal of Heredity
J. Meteorol.	Journal of Meteorology
J. Sediment. Petrol.	Journal of Sedimentary Petrology
J. Soil Water Conserv.	Journal of Soil and Water Conservation
J. Water Pollut. Control Fed.	Journal of the Water Pollution Control Federation
Mineral. Mag.	Mineralogical Magazine
Natl. Bur. Stand.	National Bureau of Standards
Nature (City location)	Nature
N.Z. J. Agric. Res.	New Zealand Journal of Agricultural Research
Nucl. Sci. Abstr.	Nuclear Science Abstracts
Phytopathology	Phytopathology
Plant Physiol.	Plant Physiology
Plant Soil	Plant and Soil
Proc. Am. Soc. Hortic. Sci.	Proceedings of the American Society for Horticultural Science
Proc. Int. Grassl. Congr., 7th	1960 Proceedings of the International Grassland 7th Congress
Proc. Int. Seed Test. Assoc.	Proceedings of the International Seed Testing Association
Proc. Natl. Acad. Sci. USA	Proceedings of the National Academy of Sciences of the United States of America
Proc. Soil Crop Sci. Soc. Fla.	Proceedings of the Soil and Crop Science Society of Florida
Residue Rev.	Residue Reviews
Science (Washington, DC)	Science
Soil Sci.	Soil Science

Soil Sci. Soc. Am. Proc.	Soil Science Society of America Proceedings
Soil Sci. Soc. Am. J.	Soil Science Society of America Journal
Soils Fert.	Soils and Fertilizers
Trans. Am. Geophys. Union	Transactions of the American Geophysical Union
Trans. Int. Congr. Soil Sci., 7th	Transactions of the 7th International Congress of Soil Science
Trans. J. Meet. Comm. 4, 5 Int. Soc. Soil Sci.	Transactions of Joint Meeting of Commissions IV & V, International Society of Soil Science
Water Resour. Res.	Water Resources Research

References

Angione, H. (ed.) 1977. The Associated Press stylebook and libel manual. Associated Press, New York.

Brickell, D.D. (ed.) 1980. International code of nomenclature of cultivated plants. Regnum Veg. 104.

Carmer, S.G., and W.M. Walker. 1982. Baby bear's dilemma: A statistical tale. Agron J. 74:122–124.

Chemical Abstracts Service. 1984. Chemical Abstracts Service source index: 1907–1984 cumulative, plus annual supplements. Chemical Abstracts Service, Columbus, OH.

Chew, V. 1976. Comparing treatment means: A compendium. HortScience 11:348–357.

Council of Biology Editors. 1983. CBE style manual. 5th ed. Council of Biology Editors, Bethesda, MD.

Crop Science Society of America. 1982. Registered field crop varieties: 1926–1981. Crop Science Society of America, Madison, WI.

Dybing, C.D. 1977. Letter from the editor (on light intensity). Crop Sci. 17:ii (March-April).

Gove, P.B. (ed.) 1964. Webster's third new international dictionary of the English language, unabridged. G. and C. Merriam Co., Springfield, MA.

Guthrie, R.L., and J.E. Witty. 1982. New designations for soil horizons and layers and the new soil survey manual. Soil Sci. Soc. Am. J. 46:443–444.

Leonard, W.H., R.M. Love, and M.E. Heath. 1968. Crop terminology today. Crop Sci. 8:257–261.

Little, T.M. 1978. If Galileo published in HortScience. HortScience 13:504–506.

Nelson, L.A., and J.O. Rawlings. 1983. Ten common misuses of statistics in agronomic research and reporting. J. Agron. Educ. 12:100–105.

Petersen, R.G. 1977. Use and misuse of multiple comparison procedures. Agron. J. 69:205–208.

Shibles, R. 1976. Committee report. Terminology pertaining to photosynthesis. Crop Sci. 16:437–439.

Soil Science Society of America, Terminology Committees. 1984. Glossary of soil science terms. rev. ed. Soil Science Society of America, Madison, WI.

Soil Survey Staff. 1975. Soil taxonomy: A basic system of soil classification for making and interpreting soil surveys. USDA-SCS. Agric. Handb. 436. U.S. Government Printing Office, Washington, DC.

Soil Survey Staff. 1978. Classification of soil series of the United States. USDA-SCS-1978 0-726-987/1626. U.S. Government Printing Office, Washington, DC.

University of Chicago Press. 1982. The Chicago manual of style. 13th ed. University of Chicago Press, Chicago.

Urdang, L. (ed.) 1972. The Random House college dictionary. Random House, New York.

U.S. Government Printing Office. 1973. Style manual. rev. ed. U.S. Government Printing Office, Washington, DC.

Weed Science Society of America. 1983. Herbicide handbook. 5th ed. Weed Science Society of America, Champaign, IL.

Other Useful Literature

American Chemical Society, Biological Abstracts and Engineering Index. 1974. Bibliographic guide for editors and authors. Chemical Abstracts Service, Columbus, OH.

American Institute of Physics. 1978. Style manual. 3rd ed., rev. D. Hathwell and A.W.K. Metzner (ed.) American Institute of Physics, New York.

American Mathematical Society. 1973. A manual for authors of mathematical papers. American Mathematical Society, Providence, RI.

American Society for Testing and Materials. 1980. ASTM Standard for Metric Practice E380-79. American Society for Testing and Materials, Philadelphia, PA.

Bates, R.L., and J.A. Jackson (ed.) 1980. Glossary of geology. 2nd ed. American Geological Institute, Alexandria, VA.

Campbell, G.S., and Jan van Schilfgaarde. 1981. Use of SI units in soil physics. J. Agron. Educ. 10:73–74.

DeBakey, L. 1976. The scientific journal: Editorial policies and practices. C.V. Mosby Co., St. Louis, MO.

Entomological Society of America. 1982. Common names of insects and related organisms. Entomological Society of America, College Park, MD.

Incoll, L.D., S.P. Long, and M.R. Ashmore. 1977. SI units in publications in plant science. Curr. Adv. Plant Sci. 28:331–342.

Little, T.M., and F.J. Hills. 1980. Statistical methods in agricultural research. John Wiley and Sons, New York.

National Bureau of Standards. 1977. International system of units. Spec. Pub. 30. U.S. Government Printing Office, Washington, DC.

O'Connor, M. 1979. The scientist as editor. John Wiley and Sons, New York.

Skillin, M.E., R.M. Gay, and other authorities [sic]. 1974. Words into type. 3rd ed. Prentice-Hall, Englewood Cliffs, NJ.

Snedecor, G.W., and W.G. Cochran. 1967. Statistical methods. 6th ed. Iowa State University Press, Ames, IA.

Steel, R.G.D., and J.H. Torrie. 1980. Principles and procedures of statistics. 2nd ed. McGraw-Hill Book Co., New York.

Strunk, W., Jr., and E.B. White. 1959. The elements of style. Macmillan Co., New York.

Terrell, E.E. Guidelines for using scientific names of plants in manuscripts. ARC, USDA Pub. PSR-26-72 (no date).

Terrell, E.E. 1977. A checklist of names for 3000 vascular plants of economic importance. USDA-ARS Agric. Handb. 505. U.S. Government Printing Office, Washington, DC.

Thien, S.J., and J.D. Oster. 1981. The international system of units and its particular application to soil chemistry. J. Agron. Educ. 10:62–70.

Vorst, J.J., L.E. Schweitzer, and V.L. Lechtenberg. 1981. International system of units (SI): Application to crop science. J. Agron. Educ. 10:70–72.

Weed Science Society of America. 1971. Composite list of weeds. Weed Sci. 20:435–476.

West, T.S. 1978. Recommendations on the usage of the terms 'equivalent' and 'normal.' Pure Appl. Chem. 50:325–338.

Woodford, F.P. (ed.) 1968. Scientific writing for graduate students. Rockefeller University Press, New York.

Index

Abbreviations, 52–54
  cultivars, 36
  federal agencies, 30
  for publication titles, 81
  in graphs, 71
  list, 53–54
  number, 58
  publication titles, 81–83
  SI units, 39, 41–43, 45, 48–51
  symbols for statistics, 40
Abstract, 23–24
  citations, 29
  in *Crops and Soils Magazine*, 12
  parts, 23
Accents, 59, 74
Acknowledgments for monographs, 16
Additional index words, 24
Administration of journals, 3
Advances in Agronomy Series reference, 32
Affect/effect, 58
Agronomy Abstracts, 15, 33
Agronomy Journal
  editorial board, 5
  invited papers, 6
  responsibilities of editorial board, 6
    associate editor, 6
    editor, 6
    managing editor, 7
    technical editor, 6
  submitting manuscripts, 3
  types of articles, 5, 6
Agronomy News, 20
Alterations to proofs, 75
American Chemical Society Handbook for Authors, 36
Analysis of variance tables, 38, 40
Appointment of editors, 5
Associate editors, 5, 6
Authors
  alterations, charges, 4
  errors in proofs, 75
  names, initials, 23, 52
  of monograph chapters, 16
  titles, 22

Base units in SI system, 41
Biology nomenclature, 35
Boards of directors, 1
Books, 15, 18, 30 (See also Monographs)

  references to, 29, 33
Brackets
  in equations, 74
  in math, 72
  in text, 56

CBE Style Manual, 35
Calendar day, 57
Capitalization
  and spelling, 54
  titles of articles in references, 30
Captions
  for figures, 28, 63
  for tables, 61
Cation exchange capacity in SI, 46
Celsius vs. Kelvin scale, 42
Chapter
  in book reference, 31
  outlines for monographs, 17
Characters, special mathematics, 72
Charges for publication, 4
Charts (see Figures)
Chemical Abstracts Service periodical titles, 30
Chemical elements, abbreviations, 52
Chemistry, nomenclature, 36
Chicago Manual of Style, 35
Chromosomes, numbering sets, 58
Code of nomenclature for cultivated plants, 35
Colon, 54
Column width of graphs, 65
Comma
  in changes of subject, 58
  in date, series, quotes, 56
  in numerals, 55
  in references, 30
Complimentary copies of monographs, 17
Composition of mixtures, percentage, 47
Compound words, 56
Computer
  graphs for publication, 71
  programs in statistical analysis, 40
Concentration, reporting in SI system, 44, 45
Conclusions in discussion section, 26
Conference reference, 31
Configuration of chemicals, 37
Consulting editors, 5
Controversy in discussion section, 26

87

Conventions and style, 35–59
Conversion factors for SI, 48–51
Coordination of editorial policies, 1
Copies
  of manuscripts, legibility, 27
  required for journals, 3
Copyright, 79
  for special publications, 19
  obtaining permission from owners, 17, 79–80
  of *Crops and Soils Magazine*, 13
  of monographs, 17
Corporate author of article reference, 32
Corrections to proofs, 75
Cost
  of preparing manuscripts for monographs, 17
  of revision to proofs, 76
Cotton, staple length, reporting, 48
Crop Science, 9
  editorial board, 9
  events, 12
  registration articles, 10
  scope, 9, 10
  submission of manuscripts, 10
Crop registration of cultivars, 36
Crop terminology, 37
Cropping photographs, 63
Crops and Soils Magazine, 11
  how to submit material, 12
  references in Society publications, 33
  review, editing, and rewriting, 13
  rewriting of articles, 13
  tables and illustrations, 13
  types of articles, 11, 12
Cultivar
  in abstract, 23
  names, 58
  nomenclature, 36

Dash, 57
Dates and time, 57
Deadlines
  for *Agronomy News*, 20
  for monograph manuscripts, 16, 17
  on proposals for special publications, 18
  on review of special publications, 20
  on revision of manuscripts, 7
Decimal, position and size in graphs, 71
Design of experiments, 39, 40
Dictionary, use of, 54
Digits, use of significant, 55
Discussion section, 25
Dissertations
  availability of, 30
  citation, 29, 31
Drafting materials, 66
Drawings (see Figures)

Editor responsibilities
  associate editor, 6
  Crops and Soils Magazine, 13
  editor-in-chief, 5
  managing editor, 6
  monographs, 14
  technical editor, 6
Editorial committee, 5–11
  duties, 16
  for books, 18
  for *Crops and Soils Magazine*, 11
  for monographs, 16
  of special publications, 19
  policies and practices, 1, 5
Effect/affect, 58
Einstein and photon flux density, 47
Electron micrographs, reproduction of, 64
Eligibility of authors, 3
Energy of soil water, in SI system, 44
Enlargement of graphs, 71
Equations (See Mathematics)
Errors
  correcting in proofs, 75
  in reference citation, 29
Exchange composition and capacity, in SI system, 44, 45
Experiments, design, 39
Exponents in SI system, 42

Feature articles for *Crops and Soils Magazine*, 11
Federal publication, 32
Field research results, reporting, 39
Figures, 63
  acceptable and nonacceptable, 69, 70
  borders in graphs, 71
  captions, 63
  estimating space for, 27
  in Materials and Methods, 25
  labeling, 71
  preparing the drawing, 66–71
  proofs, 76
  screens, 71
  size of lettering, 67–68
  thickness of ruled lines, 66
Footnotes
  for journal papers, 28
  in tables, 62
Foreign numbers and spellings, 58
Foreword for publications, 17
Format for journal articles, 21
Forms, for permission to print and copyright, 79
Forum articles for *Crops and Soils Magazine*, 11
Fractions and equations, 72

Gas concentration in SI system, 44
Glossary of soil science terms, 38

Government publication reference, 32
Graphs (see Figures)
Greek letters, 74
Halftone reproduction quality, 64
Hardiness, 58
Headings and subheadings of manuscripts, 28
Height and style of lettering, 67
Herbicides, nomenclature, 37
Honorarium for *Crops and Soils Magazine*, 11
Hortus III, 35
However, use of, 58
Hyphen, 56

Identification of plants and soils, 24, 37, 39
Illustrations (See Figures)
Index words, additional, 24
Indexing
  journals, 24
  monographs and books, 16-17
Initials of authors' names, 23
Intersociety Editorial Policy Coordination Committee, 5
Introduction of journal article, 24
Ion nomenclature, 36, 57

Journal management and procedures, 3-10
Journal of Agronomic Education
  editorial board and scope, 8
  submitting manuscripts, 3
Journal of Environmental Quality, 8
  consulting editors, 9
  copy requirements, 3, 9
  editorial board, 8
  membership not required, 9
  preparation of papers, 8
  review of manuscripts, 9
  scope of journal, 9
Journals
  article reference, 30, 33
  footnotes, 28
  format for articles, 21-26
  organization of article, 21
  title abbreviations, 81
Julian day, 57

Kelvin vs. Celsius scale, 42

Latin names
  bionomial and trinomial, 35-36
  organisms, abbreviations, 53
Latitude/longitude, 53
Le Système International d'Unités, 41
Legends, marking for figures, 63
Length
  abstract, 23
  articles for *Crops and Soils Magazine*, 12
  introduction in article, 24
  journal article, estimating, 27
  letters, 5
  notes, 4
  titles, 22
Letters
  dry transfer, 68
  height and style, 67
  symbols, accented, 74
Letters to the editor, 4
Light
  measurements and photosynthesis, 38
  SI units, 45, 47
Line drawings (See Figures)
Line-numbering of paper, 27
Literature reviews and introduction, 24

Magazine article reference, 31
Mailing
  photographs, 65
  proofs, 77
Managing editor responsibilities, 7
  journals, 7
  monographs and books, 17
Manuscripts
  acknowledging receipt, 4
  copies, 3, 27
  date of receipt, 7
  details of preparation, 27-33
  for books, 18
  for monographs, 16
  for special publications, 19-20
  format for journal articles, 21-26
  handling, 3, 7
  legibility, 27
  line-numbered paper, 27
  submitted for nonjournals, 26
  time limit on revision, 7
  typing, 27
Materials and methods, 24
Mathematics, 61, 72-74
Measurements, SI system, 41-51
Micronaire, fiber fineness reporting, 48
Mol, definition in SI, 44
Molality in SI system, 44
Monographs
  acknowledgments, 16
  complimentary copies, 17
  costs of preparing figures, 17
  distribution, 17
  duties of authors, 17
  duties of editor and editorial committee, 16
  duties of headquarters staff, 17
  feasibility committee, 16
  instructions on preparation, 15-17
  outlines, 16
  procedures for, 15

Monographs (cont.)
  promotion, 17
  references in Society publications, 33
  subject index, 16
Months, abbreviations, 52

Name/year reference system, 29
Names
  of authors, 23, 52
  of chemicals, 36
  of chemicals and crops in titles, 22
  of plants, insects, and chemicals in abstract, 23
  of special units in SI system, 41
National Cooperative Soil Survey, 38
Necrology, in *Agronomy News*, 20
Negatives, for photographs submitted, 64
Newsletter of the society, 20
Nitrogen fixation, style, 58
Nomenclature and terminology, 35–38
  biology, 35
  chemistry, 36
  crop, 37
  light measurement, photosynthesis, 38
  soil, 37–38
  statistical analysis and experimental design, 37
Non-SI units, factors for converting, 50
Normality in SI system, 44
Notes, 4
  length of abstract in, 23
Numbers and ranges, 58
Numerals, 55
  height of, in figures, 67
Numerical data, reporting, 39, 55, 61
Nutrients
  concentration in SI, 45
  elements, nomenclature, 36

Overscores, 59

Parenthesis, 56, 72
Parts per million and SI, 48
Patent reference, 32
Percentage, use in SI system, 47
Periodicals, reference to, 29
Permission
  obtaining from persons in photographs, 65
  to print and reprint, 7, 79
Personal communications, 29
Pesticides, nomenclature for, 37
Photocopies of graphs for review, 66
Photographs, 63
  cropping, 64
  for *Crops and Soils Magazine*, 13
  identifying, 65
  in *Agronomy News*, 20
  permissions for, 65
Photomicrographs, 63

Photon flux density, 47
Photosynthesis terminology, 38
Plant nutrient concentration in SI, 47
Positions available, in *Agronomy News*, 20
Postal service code, abbreviations, 52
Precision measures, reporting data, 40
Preface for publications, 16
Prefixes
  for SI units, 42
  for chemicals, 37
  use of, 56
Preparation
  drawings, 66
  manuscripts, 21
  of manuscripts for monographs, 17
  of papers for publication, 7
Printer errors in proofs, 75
Prior publication, 4
Proofs
  alterations, 75
  monograph, book, and special publication chapters, 17
  printer errors, 75
  proofreading, 75–78
  time limit for returning, 8, 77
Publications
  charges, 4
  management, 1
  of the societies, 1, 5, 8–11, 20
  title abbreviations, 81
Punctuation
  general, 55, 56
  with SI system, 43

Quotation marks, 36, 56

Ranges, 55, 58
Receipt of manuscripts
  acknowledged, 4
  date of receipt, 7
Reduction
  of graphs, 65, 66
  of photographs, 64
References, 28–33
  article with no identifiable author, 31
  availability of, 28
  conferences, 31
  dissertations, 30
  errors in citations, 29
  for reports, 31
  in discussion section, 26
  name/year vs. number, 29
  number method of citation, 29
  organization, 26
  pagination, 29
  society publications, 33
  style of, 30–33
  to literature in introductions, 24
  verification of citations, 29

Registered Field Crop Varieties: 1926–1981, 36
Replication of experiments, 39
Report reference, 31
Reprints
 journal articles, production and shipping, 8
 monograph chapters, 17
Research design, reporting, 39
Results section, 25
Retirements, in *Agronomy News*, 20
Retrieval of information, from abstract, 23
Review of manuscripts
 books, 16
 Crops and Soils Magazine articles, 13
 journal articles, 7
 monograph chapters, 16, 17
 special publications, 19
Revision
 of manuscripts, 7
 of monographs, 15
 time limit, 7
 to proof, cost, 76
Rhizobium, 59

SI System
 abbreviations, 39, 41–43, 45, 48–51
 cation exchange capacity, 46
 concentration, 44
 conversion factors, 50–51
 cotton fiber, reporting in, 48
 derived units, 41
 Einstein, 47
 energy
  of soil water, 46
  of water potential, 46
 exchange composition, 46
 exponents, 42
 light, 47
 molarity, 44
 non-SI units, 42
 normality, 44
 parts per million, 48
 percentage, use in, 47
 photon flux density, 47
 preferred and alternate units, 48–50
 prefixes for units, 42
 punctuation, 43
 soil water energy, 46
 symbols for units, 41
 time, 44
 units, 41, 42
 water potential energy, 46
Scale of photomicrographs, 63
Series
 articles, 22, 23
 measurements, 53
 pictures, 64
 titles, 22

Signs
 integral and limits, 73
 radical, 73
Slant line, 43
Slides, use of as figures, 63
Society publications reference, 33
Soil
 description in field experiments, 37
 identification, 24, 37
 nomenclature, 37–38
 reporting water in SI system, 46
 terminology, 38–39
 terms, glossary of, 38
Soil Science Society of America Journal, 10
 copy requirements, 3, 10
 editorial board, 10
 publication policy, 10
 review of manuscripts, 10
 scope, 10
 terminology, 38
Space, sufficient use of in graphs, 71
Special publications
 guidelines for, 18–20
 references in Society publications, 33
 sponsorship of, 19
Speculation in discussion section, 26
Spelling and capitalization, 54
Staple length, of cotton, 48
State publications, 32
Statistics
 abbreviations used in, 40
 comparisons among treatments, 38
 nomenclature, 39
Submission
 manuscripts to journals, 3
 of articles to *Crops and Soils Magazine*, 12
 of monograph and book manuscripts, 16
Subscripts, 59
Suffixes, use of, 56
Superscripts, 59
Symbols
 and abbreviations, 52
 for base units in SI system, 41
 for chemistry, 36
 for SI units, 41, 42
 in graphs, 65
 in tables, 62
Symposium
 publication, 18
 reference, 31

Tables, 61
 and figures, 26
 estimating space for, 28
 in Materials and Methods, 25
 proofs, 76
 sample, 62

Technical editors, 5, 6
Terminology
  for soil reporting, 38
  in biology, 35
  in crop science, 37
  in special fields, 37
That/which, 59
Thesis
  citation, 29
  in *Agronomy News*, 20
Time and dates, 57
  use of time in SI system, 44
Time limit
  for returning proofs, 8
  for review of special publications, 20
  on revision of journal manuscripts, 7
Title
  abbreviations, of publications, 81
  and authors, 22
  and index words, 24
  capitalization in references, 30
Tonne in SI system, 44
Transactions reference, 31
Translation reference, 31
Treatment design, reporting, 39
Typescript
  estimating, 27
  letters for graphs, 71
  manuscripts, 27
  tables, 62

U.S. Government Printing Office Style Manual, 35
Units
  and conversion factors in SI, 48
  in SI system, 41, 42
  in tables, 61
  non-SI units, 43
  of measure, lettering in, 67
  of measurement in monographs, 17
  of measurement, abbreviation, 52
  preferred and acceptable in SI, 45, 48
University Microfilms, 30
Usage, 57

Validity of results, 39

Was, were, 59
Water
  content in SI, 45
  potential in SI system, 44–46
Which/that, 59
Width of publication, photos for, 64
Word limit for abstracts, 23
Workshop proceedings reference, 31